Agronomic and Ecological Effects of Plastic Film Mulching in Dryland Farming

地膜覆盖在旱地农业中的农学和生态效应

◎ 陈保青　董雯怡　严昌荣　著

中国农业科学技术出版社

图书在版编目(CIP)数据

地膜覆盖在旱地农业中的农学和生态效应：Agronomic and Ecological Effects of Plastic Film Mulching in Dryland Farming：英文／陈保青，董雯怡，严昌荣著．—北京：中国农业科学技术出版社，2020.10
ISBN 978-7-5116-4999-7

Ⅰ.①地… Ⅱ.①陈…②董…③严… Ⅲ.①地膜覆盖-研究-英文 Ⅳ.①S626.4

中国版本图书馆 CIP 数据核字(2020)第 167491 号

| 责任编辑 | 金　迪　崔改泵 |
| 责任校对 | 贾海霞 |

出版者	中国农业科学技术出版社
	北京市中关村南大街 12 号　邮编：100081
电　话	(010) 82109194 (编辑室)　(010) 82109702 (发行部)
	(010) 82109709 (读者服务部)
传　真	(010) 82109698
网　址	http://www.castp.cn
经销者	各地新华书店
印刷者	北京建宏印刷有限公司
开　本	787 mm×1 092 mm　1/16
印　张	8.75
字　数	219 千字
版　次	2020 年 10 月第 1 版　2020 年 10 月第 1 次印刷
定　价	68.00 元

版权所有·翻印必究

Agronomic and Ecological Effects of Plastic Film Mulching in Dryland Farming

著者名单

主著：陈保青　董雯怡　严昌荣
著者：刘恩科　刘　勤　刘　秀
　　　杨　潇　白重九　王尚文
　　　Sarah Garré　Shahar Baram

Preface

How to feed a growing population with less environmental cost is one of the most important issues for global agriculture development. In dry-land, which covers about 41% of the Earth's surface, plastic mulching technology is one of the most widely used technologies to overcome water shortages and improve crop yields because of their ability to improve soil temperature, conserve water, enable earlier harvest and inhibit weed growth. It was estimated that the plastic mulching could improve the crop yield by 20-60%. The high improvement in crop yield, the convenient methods, and the good economic returns caused the use of plastic mulching to quickly spread in dry-land farming systems worldwide.

However, during the past 10 years, more and more negative environmental effects of plastic mulching have been found. The application of plastic mulching not only can improve the risks of soil degradation, greenhouse gas emissions, and solution leaching by modifying the micro-environment during the mulching period, but can also produce plastic residue pollution, toxicants, and micro/nano plastic pollution when the plastic film is not retrieved well. Development of environment-friendly mulching technology has become an important and urgent issue for the development of global drylands.

In the past few years, a series of researches were carried out by us to evaluate the effects of plastic mulching on crop yield, soil micro-climate conditions, soil microbial community, greenhouse gas emission and economic benefits, and at the same time, new sustainable plastic mulching technology was proposed. This book is a summary of our researches, and we hope our results can provide fundamental reference for future researches on this subject.

Content

1 Application of Plastic Mulching in Dryland

2 Effects of Plastic Mulching on Soil Water Content, Water Use Efficiency and Economy Benefits
- 2.1 Abstract ……………………………………………………………… 17
- 2.2 Introduction …………………………………………………………… 17
- 2.3 Materials and methods ………………………………………………… 19
- 2.4 Results ………………………………………………………………… 25
- 2.5 Discussion …………………………………………………………… 30
- 2.6 Conclusion …………………………………………………………… 34
- 2.7 Reference ……………………………………………………………… 34

3 Modeling of Soil Water Dynamic and Field-scale Spatial Variation of Soil Water Content
- 3.1 Abstract ……………………………………………………………… 41
- 3.2 Introduction …………………………………………………………… 41
- 3.3 Materials and methods ………………………………………………… 43
- 3.4 Results ………………………………………………………………… 50
- 3.5 Discussion …………………………………………………………… 54
- 3.6 Conclusion …………………………………………………………… 56
- 3.7 Reference ……………………………………………………………… 56

4 Field-scale Spatial Variation of Soil Moisture Fluctuation
- 4.1 Abstract ……………………………………………………………… 63
- 4.2 Introduction …………………………………………………………… 63
- 4.3 Materials and Methods ………………………………………………… 65
- 4.4 Results and discussion ………………………………………………… 69
- 4.5 Conclusion …………………………………………………………… 76
- 4.6 Reference ……………………………………………………………… 76

5 Effects of Plastic Mulching on Soil Microbial Community
- 5.1 Abstract ……………………………………………………………… 83

5.2	Introduction	83
5.3	Materials and methods	85
5.4	Results	89
5.5	Discussion	99
5.6	Conclusion	103
5.7	Reference	103

6 Effects of Plastic Mulching on Carbon Footprint

6.1	Abstract	111
6.2	Introduction	111
6.3	Materials and methods	113
6.4	Results and discussion	118
6.5	Conclusions	124
6.6	Reference	124

1
Application of Plastic Mulching in Dryland

Drylands include arid, semi-arid, and dry sub-humid ecosystems characterized by low and irregular rainfall and high evapotranspiration. Drylands are subject to cyclical droughts, cover about 41% of the Earth's surface (Figure 1-1), and are home to more than 30% of the total global population (Reynolds et al., 2007). Water shortages in drylands lead to agricultural problems, such as a lower crop yield and a high drought risk (Hazell and Hess, 2010), societal problems, such as poverty and hunger, and environmental issues, such as soil degradation and desertification. Improving crop yields, reducing hunger, alleviating poverty, and controlling and preventing environmental degradation are important goals for the development of dryland agriculture (ICARDA, 2016).

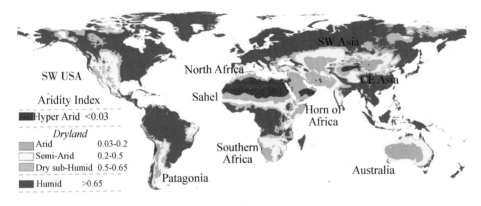

Figure 1-1 Global distribution of dryland

Many agricultural practices, such as conservation tillage, breeding crops with high drought resistance, and creating anti-transpirants, water-retaining agents, and mulching technology, have been developed to overcome water shortages and improve crop yields in drylands (Cattivelli et al., 2008; Mallikarjunarao et al., 2015; Van Wie et al., 2013; Qin et al., 2015). Among them, plastic mulching technology (PM) is one of most widely used. Unlike the plastic used in tunnel cultivation, the plastic used in PM technology is relatively thin and directly covers the soil surface without covering the crop canopy. The history of PM technology dates back to the mid-1950s, when Dr. Emery M. Emmert from the University of Kentucky was the first to use polyethylene (PE) as a greenhouse film and detailed the principles of plastic technology (Emmert, 1957). Up until now, PM practices have been adopted in many countries because of their ability to improve soil temperature and crop yield, enable earlier harvest, inhibit weed growth, and conserve water. In Europe, the application of PM has reached 427×10^3 ha, which is four times larger than the area occupied by greenhouses and six times larger than the area occupied by low tunnels (Scarascia-Mugnozza et al., 2011). Especially in some Mediterranean countries (e.g., Spain, France, Italy), PM is effective in reducing drought stress and enabling crop production (Scarascia-Mugnozza et al., 2011). In Europe, PM is mainly used for the production of flowers, ornamentals, and vegetables, such as tomatoes, eggplants, peppers, cucumbers, asparagus, melons, watermelon, and strawberries (Scarascia-Mugnozza et al., 2011). In China, the area has increased more than 150-folds from 1982 to 2016 (Figure 1-2), and, in 2016, it reached 18401×10^3 ha, which is equal to about 14% of the total farmland of China (NBSC, 2017). In 1980s, PM led to a tremendous increase in the yield of peanuts, vegetables, and cotton, and it was the start of what was called the white revolution in China (Hu et al., 1995; Liu et al., 2014). Currently, PM has been widely applied in the cultivation of maize, wheat, cotton, rice, soybeans, potatoes, vegetables, and fruit (Figure 1-3). White (or transparent) and black plastic films are the two most widely used films. In the Mediterranean region of Europe, black film is preferred because of its ability to control weed growth. In Northeast China, which is characterized by its high latitude, and in Northwest China, which is characterized by its high altitude, white plastic film is more popular because of its ability to overcome temperature stress (Tarara, 2000).

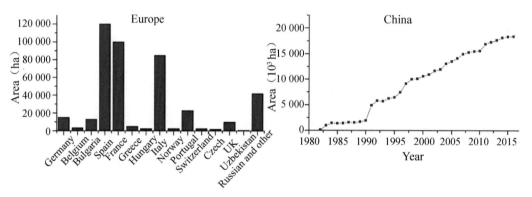

Figure 1-2 Application area of plastic mulching in European countries and China

Figure 1-3 Application of plastic mulching in the cultivation of different crops

The increase in the human population has led to a huge food demand (Godfray et al., 2010), and PM can increase food production, even with the limited availability of arable land. The improvement that PM enables on crop yields results from its modification of the soil's microclimate (Tarara, 2000). Previous research indicated that PM can increase the soil temperature through three pathways. In the first, the plastic film usually has a relatively high shortwave transmittance coupled with a relatively low longwave transmittance, which can heat the soil by transmitting and/or absorbing shortwave radiation while preventing the loss of emitted radiant energy in the longwave spectrum (Ham et al., 1993). In the second, when plastic film is placed loosely on the soil surface, an insulating air gap is established, and then greater heat storage or less heat loss occurs. Compared to the first pathway, this pathway is usually more important for warming soil under plastic mulch (Ham et al., 1993; Ham et al., 1994). In the third, evaporation is reduced under PM, and this may subsequently reduce the latent heat flux of the surface soil. The influence of plastic on soil moisture, which is related to reduce evaporation fluxes, improves water availability and enables plant growth (Allen et al., 1998; Gong et al., 2017). Thus, the application of PM can alleviate temperature and water stress and enable crop growth in cold or dry regions, thus improving crop yield (Figure 1-4). According to a review that compared the crop yields between soils treated with and without PM in China, when PM was adopted, grain and cash crop yields were increased by 20-35% and 20-60%, respectively (Liu et al., 2014).

Figure 1-4 Contrasting crop development in no-mulch and plastic mulch field

Although plastic mulching has great potential to improve the crop yield, its negative influences on environment are widely debated. Previous researches suggested application of plastic mulching could influence soil heath, greenhouse gas emission and nutrient transport through changing micro-environment conditions during mulching period and effects of plastic residues (Figure 1-5).

Better soil moisture and temperature can improve the crop photosynthesis and subsequently improve the input of soil organic matter through plant residue and root excreta (Ali et al., 2018; Yong et al., 2016; An et al., 2015); however, because the improved soil moisture

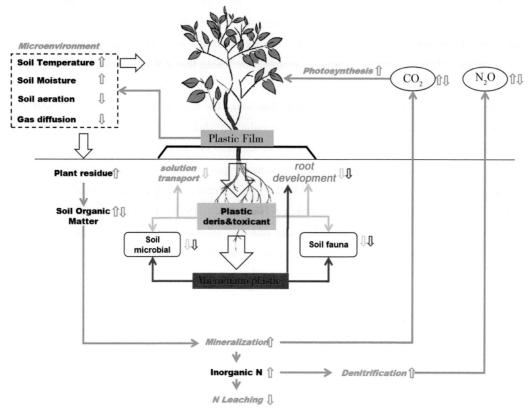

Figure 1-5　Overview of the main environment risks of plastic mulching technology

and temperature tend to improve the soil microbial community and activity, the consumption of soil organic matter by microbials is also improved, and the mineralization rate of soil organic matter is improved (Dong et al., 2017; Liu et al., 2019; Hai et al., 2015). As such, the application of plastic film was found to reduce the soil organic matter in some studies (Ma et al., 2018; Li et al., 2004; Zhou et al., 2012) and had positive or no obvious effects in other studies (Zhang et al., 2017; Wang et al., 2016; Luo et al., 2015). As soil organic matter is crucial to sustaining soil structure and fertility, in sites where plastic mulching induces the reduction of soil organic matter, the soil will face risks of decline in soil structural stability and productivity. Recent studies also suggested that application of plastic mulching has risks to improve the GHG emission. In some cases, it was found that the application can improve CO_2 efflux based on results of improved microbial activity and changed rates of soil organic matter decomposition (Zhang et al., 2015; Yu et al., 2018; Chen et al., 2017; Liu et al., 2016); however, other studies showed that the application of PM can reduce CO_2 efflux due to the limitations of plastic mulching on the diffusion of gas (Li et al., 2012; Li et al., 2011; Okuda et al., 2007). At the same time, some researches also suggested that the application of plastic mulching improved N_2O emissions due to lower aeration (Kim et al., 2017;

Cuello et al., 2015; Nan et al., 2016), while Berger et al. (2013) suggested that the application of plastic mulching decreased N_2O emissions based on the extremely low soil moisture at their sites. Those studies suggested that the effects of plastic mulching on soil degradation and GHG emission varied from site to site, and there was still lacking of knowledge to determine whether it was safe for a specific farmland.

Polyethylene (PE) is a commonly used polymer for the production of plastic film. It possess a carbon backbone that is resistant to hydrolytic and enzymatic degradation. The lifetime of PE can be as long as hundreds of years, and when it is incorporated into the soil, it is almost impossible to be assimilated and mineralized by the microbes. If the PE film is not retrieved after usage, it will accumulate in the soil and lead to several negative effects (Figure 1-6), such as retarding crop root growth (Liu et al., 2014), reducing the continuity of soil porosity, and impeding the transport of soil water and nutrients (Jiang et al., 2017), inhibiting soil microbial and enzyme activities and thus leading to a decline in soil nutrient status (Zhang et al., 2017). Generally, the accumulation of plastic film in soil has negligible effects on crop yields at low doses, while the yield decline risk increases with the plastic film residue amount in the soil. In a report from Gao et al. (2018), no significant effect on crop yield was found for residual plastic between 0 and 240 kg·ha^{-1}, but the yield decreased significantly with residual plastic film >240 kg·ha^{-1}. Moreover, plastic residue in the soil can form micro (100 nm to 5 mm) and nano (<100 nm) sized particles with photo-and thermo-oxidative degradation and biodegradation by microorganisms after a long period of environmental exposure and oxidation (Ng et al., 2018), which can be taken up by organisms and causes the problems of bioaccumulation and biomagnification (Galloway et al., 2017; Rochman et al., 2016). Ingestion of microplastics by organisms leads to weight loss and even death (Huerta Lwanga et al., 2018; Cao et al., 2017; Rodriguez-Seijo et al., 2017), while nanoplastics are suggested to be more hazardous because they can permeate biological membranes and can harm soil microbiomes and plants (Ng et al., 2018). In addition, previous researches suggested that some kinds of ploymers and additives used for plastic film manufacturing can produce exotoxicity. The ecotoxicity of plastic films can be tested by different methods, which mainly include the determination of seed germination and the growth of plants, acute toxicity on earthworms, acute toxicity on Daphnia, and effects on microbial nitrification, bacteria, crustaceans, etc. (Sforzini et al., 2016; Chae and An, 2018). Additives are chemically or physically active materials used during plastic film processing to create new chemical structures or to modify physical characteristics, such as mechanical properties, optical characteristics, and oxidation resistance (Kyrikou and Briassoulis, 2007). Additives used for the processing of plastic film mainly include antioxidants, light stabilizers, plasticizers, photosensitizers, coupling agent, antiblocking agents, chain extenders, talc, etc. Most additives are non-toxic; however, plasticizers and some antioxidants, such as substituted phenylamine antioxidants, are believed to be potentially carcinogenic and mutagenic (Matsumoto et al., 2008; Prosser et al., 2017).

Figure 1-6　Plastic residue pollution in arable lands due
to long-term plastic film mulching

Therefore, development of sustainable plastic mulching system is an important issue for dryland farming. This book is a summary of our researches on the subject of development of sustainable plastic mulching systems in the past five years (2015-2019), and we hope our results can provide fundamental reference for future researches on this subject.

Currently, scientists are searching for approaches to solve these problems, and these works mainly include the development of biodegradable film and the reduction of film using film recycling machines. At the current technical level, the cost of biodegradable film is around two to five times higher than that of PE film. This is a dominant factor that restricts its popularization. Moreover, biodegradable film has obvious disadvantages in terms of moisture conservation and temperature improvement, and this may limit its application in dry areas. The development of machinery for residual film recovery is another measure to control plastic film pollution. However, its collecting efficiency for thin film still needs to be improved. Moreover, using this machinery results in additional operation costs. How to reduce the plastic residue pollution meanwhile improve economic benefits in agriculture production is one of most difficult problem for the development of sustainable plastic mulching system. It is highly related with the willingness of farmers to accept new technologies. In the first part of our research (Chapter 2), we proposed a kind of new plastic mulching pattern, namely, "one film for 2 years" system (PM2), to solve this problem, and the crop yield, water use efficiency, and economic benefits were compared among no mulch, traditional plastic mulching, and "one film for 2 years" system.

Many environmental risks caused by plastic mulching are highly related with the effects of plastic mulching on soil moisture and temperature and understanding the influences of plastic mulching on soil moisture and temperature will provide strong information support for the understanding of biogeochemical processes, such as the mineralization of soil organic matter, nitrification, denitrification, etc., and then improve management practices, such as the sowing position, fertilization position, bare strip, and mulched strip. Although many publications exist containing the words plastic mulch and soil water, a 2D understanding of soil water fluxes is only present in a few articles. Point-scale soil moisture measurement methods, such as gravimetric measurements on soil samples, neutron probes, time-domain reflectometry, and dielectric constant sensors, have been widely used to study the effect of plastic mulching on soil moisture. However, these methods only provide information about a relatively small volume of soil, often limited to a few centimeters. Using a network of point-scale sensors can be expensive and impractical, and, normally, the number of sensors or probes cannot be too high because installing such sensors or probes disturbs the soil. Thus, these approaches are usually characterized by a low spatial resolution, and, when the soil moisture shows high spatial variations, the information obtained using these approaches is insufficiently representative of water processes and could lead to a large degree of uncertainty and errors. Spatial variations in the soil water dynamics at the field scale can be addressed using 2D modeling or geophysical techniques. During the past decades, many soil hydraulic models have been developed at different scales, such as WaSim-ETH, the Community Land Model, SiSPAT-Isotope, and Hydrus. Among them, Hydrus has been widely used at the field scale and can be used to predict 2D or three-dimensional (3D) soil water dynamics by applying different types of boundary conditions. However, Hydrus 2D has not been used to predict the soil water dynamics in rainfed dryland PM fields. Geophysical techniques, such as ground-penetrating radar, electromagnetic induction, and ERT, can be used to obtain spatial information because they are usually noninvasive and able to obtain a large number of measurements rapidly. Surface ERT is a noninvasive technology that can be used to map the 2D or 3D distribution of electrical resistivity, which is linked to SWC through petrophysical relationships. It has been increasingly used to study the spatial variations in soil water at the field scale; however, it still has not been applied to address the spatial variations in soil water in PM fields. As such, in the second and third part of our research (Chapter 3 and Chapter 4), we developed a kind of simulation methods which could obtain soil moisture and temperature information and develop methods to use electrical resistivity tomography (ERT) technology to obtain the soil moisture information in field with plastic mulching, and based on the developed simulation methods and ERT technologies, we analyzed the spatial and temporal variation of soil moisture and temperature in field with plastic mulching.

The soil microbial community plays a crucial role in nutrient cycling, maintenance of soil structure and its diversity is a sensitive indicator of soil quality that can reflect subtle changes and provide information for evaluation of soil function. The determination of factors that influence microbial community composition under field conditions has significantly increased our understanding of how management affects crop quality, disease ecology, and biogeochemical cycling, and research on the effects of soil management on soil microbial communities has become a fundamental aspect of sustainable agriculture. In the fourth part of our research (Chapter 5), Illumina Hiseq sequencer was employed to compare the soil bacterial and fungal communities in no mulch field and plastic mulched field. Carbon footprint, defined as the total amount of carbon emissions caused directly and indirectly by an activity, or the emissions that accumulate over the life cycle of a product from cradle to grave, has become an important method to systematically evaluate the carbon emissions caused by artificial factors in the agricultural production process. Carbon footprint evaluation for crop products provides a reliable way to recognize the key sources of GHG emissions in different cultivation systems and to quantify the influences of agricultural practices and other environmental factors. As such, it provides the theoretical basis for optimizing agricultural management practices to reduce GHG emissions and developing low-carbon agriculture. In the last part of our research (Chapter 6), we evaluated the carbon footprint for three maize cultivation systems-no mulch system, an annual plastic mulching system and "one film for 2 years" system.

Our researches were supported by the National Natural Science Foundation of China (grant numbers: 31871575, 41601328, 31901477, 3191101554), and we thank the scholarship "Erasmus+International Credit Mobility" awarded by the University of Liège on behalf of European Commission, for the stay in Gembloux Agro-Bio Tech, and postdoctoral fellowships provided by Ministry of Agriculture and Rural Development of Israel, for the stay in Volcani Center. Our deepest gratitude goes foremost to Professor Garré Sarah (Universtiy of Liege, Belgium) and Professor Shahar Baram (Volcani Center, Israel), and the Central Public Interest Scientific Institute Basal Research Fund (No: BSRF201909). We also would like to express our heartfelt gratitude to Professor Gilles Colinet, Aurore Degré, Benjamin Dumont, Minggang Xu, and Yanqing Zhang who have instructed and helped us a lot in the past five years.

Reference

Ali S., Xu Y., Jia Q., et al, 2018. Interactive effects of plastic film mulching with supplemental irrigation on winter wheat photosynthesis, chlorophyll fluorescence and yield under simulated precipitation conditions [J]. Agric. Water. Manage., 207: 1-14.

An T., Schaeffer S., Li S., et al, 2015. Carbon fluxes from plants to soil and dynamics of microbial immobilization under plastic film mulching and fertilizer application using 13 C pulse-labeling [J]. Soil Biol. Biochem., 80: 53-61.

Berger S., Kim Y., Kettering J., et al, 2013. Plastic mulching in agriculture-friend or foe of n2o emissions? [J]. Agric. Ecosyst. Environ., 167: 43-51.

Cao D., Xiao W., Luo X., et al, 2017. Effects of polystyrene microplastics on the fitness of earthworms in an agricultural soil [J]. IOP Conf. Ser. Earth. Environ. Sci., 61: 12148.

Cattivelli L., Rizza F., Badeck F. W., et al, 2008. Drought tolerance improvement in crop plants: an integrated view from breeding to genomics [J]. Field Crop. Res., 105 (1-2): 1-14.

Chae Y., and An Y. J, 2018. Current research trends on plastic pollution and ecological impacts on the soil ecosystem: A review [J]. Environ Pollut, 240: 387-395.

Chen H., Liu J., Zhang A., et al, 2017. Effects of straw and plastic film mulching on greenhouse gas emissions in Loess Plateau, China: a field study of 2 consecutive wheat-maize rotation cycles [J]. Sci. Total Environ., 579: 814-824.

Cuello J. P., Hwang H. Y., Gutierrez J., et al, 2015. Impact of plastic film mulching on increasing greenhouse gas emissions in temperate upland soil during maize cultivation [J]. Appl. Soil Ecol., 91: 48-57.

Dong W., Si P., Liu E., et al, 2017. Influence of film mulching on soil microbial community in a rainfed region of northeastern China [J]. Sci. Rep., 7 (1): 8468.

Emmert E. M, 1957. Black polyethylene for mulching vegetables [J]. Proc. Amer. Soc. Hort.Sci., 69: 464-469.

Galloway T. S., Cole M., and Lewis C, 2017. Interactions of microplastic debris throughout the marine ecosystem [J]. Nat. Ecol. Evol., 1: 116.

Gao H., Yan C., Liu Q., et al, 2018. Effects of plastic mulching and plastic residue on agricultural production: A meta-analysis [J]. Sci. Total Environ., 651: 484-492.

Godfray H. C., Beddington J. R., Crute I. R, et al, 2010. Food security: the challenge of feeding 9 billion people [J]. Science, 327: 812-818.

Hai L., Li X. G., Liu X., et al, 2015 Plastic mulch stimulates nitrogen mineralization in urea-amended soils in a semiarid environment [J]. Agron. J., 107 (3): 921-930.

Ham J. M., and Kluitenberg G. J, 1994. Modeling the effect of mulch optical properties and mulch-soil contact resistance on soil heating under plastic mulch culture [J]. Agric. For. Meteorol., 71 (3-4): 403-424.

Ham J. M., Kluitenberg G. J., and Lamont W. J, 1993. Optical properties of plastic mulches affect the field temperature regime [J]. J. Amer. Soc. Hort. Sci., 118 (2): 188-193.

Hazell P. B. R., and Hess U, 2010. Drought insurance for agricultural development and food security in dryland areas [J]. Food Secur., 2: 395-405.

Hu W., Duan S. F., and Song Q. W, 1995. High yield technology for groundnut [J]. Int. Arachis. Newsl., 15: 1-22.

Huerta Lwanga E., Thapa B., Yang X., et al, 2018. Decay of low-density polyethylene by bacteria extracted from earthworm's guts: a potential for soil restoration [J]. Sci. Total. Environ., 624: 753-757.

ICARDA (The International Center for Agricultural Research in the Dry Area). 2016. Brief ICARDA Strategy 2017-2026 [M]. http://www.icarda.org/sites/default/files/ICARDA_STRATEGY_BRIEF.pdf.

Jiang X. J., Liu W., Wang E., et al, 2017. Residual plastic mulch fragments effects on soil physical properties and water flow behavior in the Minqin Oasis, northwestern China

[J]. Soil Till. Res., 166: 100-107.

Kim G. W., Das S., Hwang H. Y., et al, 2017. Nitrous oxide emissions from soils amended by cover-crops and under plastic film mulching: fluxes, emission factors and yield-scaled emissions [J]. Atmos. Environ., 152: 377-388.

Kyrikou I. and Briassoulis D, 2007. Biodegradation of Agricultural Plastic Films: A Critical Review [J]. J. Polym. Environ., 15 (3): 227-227.

Li F., Song Q., Jjemba P., et al, 2004. Dynamics of soil microbial biomass c and soil fertility in cropland mulched with plastic film in a semiarid agro-ecosystem [J]. Soil Biol. Biochem., 36 (11), 1893-1902.

Li Z. G., Zhang R. H., Wang X. J., et al, 2012. Growing season carbon dioxide exchange in flooded non-mulching and non-flooded mulching cotton [J]. PLOS ONE 7 (11): e50760.

Li Z. G., Zhang R. H., Wang X. J., et al, 2011. Carbon dioxide fluxes and concentrations in a cotton field in Northwestern China: effects of plastic mulching and drip irrigation [J]. Pedosphere, 21 (2): 178-185.

Liu E. K., He W. Q., and Yan C. R, 2014. "White revolution' to' white pollution" -agricultural plastic film mulch in China [J]. Environ. Res. Lett., 9 (9): 3.

Liu Q., Chen Y., Li W., et al, 2016. Plastic-film mulching and urea types affect soil CO_2 emissions and grain yield in spring maize on the loess plateau, china [J]. Sci. Rep., 6: 28150.

Liu X., Dong W., Si P., et al, 2019. Linkage between soil organic carbon and the utilization of soil microbial carbon under plastic film mulching in a semi-arid agroecosystem in China [J]. Arch. Agron. Soil Sci., doi:10. 1080/03650340. 2019. 1578346.

Luo S., Zhu L., Liu J., et al, 2015. Sensitivity of soil organic carbon stocks and fractions to soil surface mulching in semiarid farmland [J]. Eur. J. Soil Biol., 67: 35-42.

Ma D., Chen L., Qu H., et al, 2018. Impacts of plastic film mulching on crop yields, soil water, nitrate, and organic carbon in Northwestern China: A meta-analysis [J]. Agric. Water Manage., 202: 166-173.

Mallikarjunarao K., Pradhan R., and Das R. K, 2015. Dry land techniques for vegetable production in India-a review [J]. Agr. Review, 36 (3): 227-234.

Matsumoto M., Hirata-Koizumi M., and Ema M, 2008. Potential adverse effects of phthalic acid esters on human health: a review of recent studies on reproduction [J]. Regul. Toxicol. Pharm., 50 (1): 37-49.

Nan W. G., Yue S. C., Huang H. Z., et al, 2016. Effects of plastic film mulching on soil greenhouse gases (CO_2, CH_4 and N_2O) concentration within soil profiles in maize fields on the Loess Plateau, China [J]. J. Integr. Agric., 15 (2): 451-464.

NBSC (National Bureau of Statistics of China), 2017. China statistical yearbook 2017 [M]. Beijing: China Statistics Press.

Ng E. L., Lwanga E. H., Eldridge S. M., et al, 2018. An overview of microplastic and nanoplastic pollution in agroecosystems [J]. Sci. Total Environ., 627: 1377-1388.

Okuda H., Noda K., Sawamoto T., et al, 2007. Emission of N_2O and CO_2 and uptake of CH_4 in soil from a Satsuma mandarin orchard under mulching cultivation in central Japan [J]. J. Jpn. Soc. Hortic. Sci., 76 (4): 279-287.

Prosser R. S., Parrott J. L., Galicia M., et al, 2017. Toxicity of sediment-associated substituted phenylamine antioxidants on the early life stages of Pimephales promelas and a characterization of effects on freshwater organisms [J]. Environ. Toxicol. Chem., 36 (10): 2730-2738.

Qin S., Yeboah S., Xu X., et al, 2017. Analysis on fungal diversity in rhizosphere soil of continuous cropping potato subjected to different furrow-ridge mulching managements [J]. Front. Microbiol., 8: 845.

Reynolds J. F., Smith D. M. S., Lambin E. F, et al, 2007. Global desertification: building a science for dryland development [J]. Science, 316 (5826): 847-851.

Rodriguez-Seijo A., Lourenço J., Rocha-Santos T. A. P., et al, 2017. Histopathological and molecular effects of microplastics in Eisenia andrei Bouché [J]. Environ. Pollut., 220: 495-503.

Scarascia-Mugnozza G., Sica C., and Russo G, 2011. Plastic materials in european agriculture: actual use and perspectives [J]. J. Agric. Eng., 42 (3): 15-28.

Sforzini S., Oliveri L., Chinaglia S., et al, 2016. Application of biotests for the determination of soil ecotoxicity after exposure to biodegradable plastics [J]. Front. Environ. Sci., 4: 68.

Tarara J. M, 2000. Microclimate modification with plastic mulch [J]. HortScience, 35 (2): 169-180.

van Wie J. B., Adam J. C., and Ullman J. L, 2013. Conservation tillage in dryland agriculture impacts watershed hydrology [J]. J. Hydrol., 483: 26-38.

Wang Y. P., Li X. G., Fu T., et al, 2016. Multi-site assessment of the effects of plastic-film mulch on the soil organic carbon balance in semiarid areas of China [J]. Agric. For. Meteorol., 228: 42-51.

Yong P. W., Xiao G. L., Fu T., et al, 2016. Multi-site assessment of the effects of plastic-film mulch on the soil organic carbon balance in semiarid areas of china [J]. Agric. For. Meteorol., 228-229: 42-51.

Yu Y., Tao H., Yao H., et al, 2018. Assessment of the effect of plastic mulching on soil respiration in the arid agricultural region of china under future climate scenarios [J]. Agric. For. Meteorol., 256: 1-9.

Zhang D., Liu H. B., Zhongming M. A., et al, 2017. Effect of residual plastic film on soil nutrient contents and microbial characteristics in the farmland [J]. Scientia Agricultura Sinica., 50 (2): 310-319.

Zhang F., Li M., Qi J., et al, 2015. Plastic film mulching increases soil respiration in ridge-furrow maize management [J]. Arid. Land Res. Manage, 29 (4): 432-453.

Zhang F., Zhang W., Li M., et al, 2017. Does long-term plastic film mulching really decrease sequestration of organic carbon in soil in the loess plateau? [J]. Eur. J. Agron., 89: 53-60.

Zhou L. M., Jin S. L., Liu C. A., et al, 2012. Ridge-furrow and plastic-mulching tillage enhances maize-soil interactions: opportunities and challenges in a semiarid agroecosystem [J]. Field Crop. Res., 126 (1): 181-188.

2

Effects of Plastic Mulching on Soil Water Content, Water Use Efficiency and Economy Benefits

2.1 Abstract

"One film for 2 years" (PM2) has been proposed as a practice to control the residual film pollution; however, its effects on grain-yield, water-use-efficiency and cost-benefit balance in dryland spring maize production have still not been systematically explored. In this study, we compared the performance of PM2 with the annual film replacement treatment (PM) and no mulch treatment (NM) on the Loess Plateau in 2015-2016. Our results indicated the following: ①PM2 was effective at improving the topsoil moisture (0-20 cm) at sowing time and at seedling stage, but there was no significant influence on soil water storage, seasonal average soil moisture or evapotranspiration; ②PM2 induced significantly higher cumulative soil temperatures compared to NM, and there was no significant difference between PM2 and PM; ③No significant differences were identified in grain-yield and water-use-efficiency between PM and PM2, and compared to NM, they were improved by 16.3% and 15.5%, respectively; ④Because of lower cost of plastic film, tillage, film deposit and remove in PM2, economic profits improved by 12% and 81% compared to PM and NM in scenario1 (without film collection), and 21% and 70% in scenario 2 (with film collection but without selling plastic residues), and 17% and 74% in scenario 3 (with film collection and selling plastic residues), respectively. This research suggested that PM2 was effective at alleviating the spring drought and was beneficial in reducing poverty in dryland.

Keywords: Plastic mulching; Soil water; Maize yield; Water use efficiency; Cost-benefit.

2.2 Introduction

Achieving high yields on existing croplands with less impact on the environment is one of the most important issues for agricultural sustainable development, and this challenge requires changes in the way food is produced (Tilman et al., 2011; Godfray et al., 2010). Plastic mulching is important for crop production in China, and from 1991 to 2011, there has been a four-fold increase in plastic mulch use (National Bureau of Statistics of China, 2014), which has generated important improvements in crop production (Liu et al., 2014; Qin et al., 2015). However, accumulation of plastic residues in soil is becoming increasingly serious, and a typical survey in China demonstrates that the residual amount in soil has reached 71.9-259.1 kg·ha^{-1} (Yan et al., 2014). It poses a direct threat to soil health and crop production (Guo et al., 2016; Dong et al., 2013; Niu et al., 2016) and also leads to high loads of phthalate esters in agricultural soils (Chen et al., 2013).

Development of film use frequency reduction, biodegradable film and machinery designed for residual film recovery are three possible ways to control plastic film pollution (Yan et al., 2014; Liu et al., 2014). Using one film for two or more years ("one film for 2 years" or "one film for multiple years") is one of the technologies used to reduce film use frequency. Current plastic mulching technology has been characterized by annually replacing film, and the residual film is usually directly incorporated into the soil during tillage because of its low recycling value. In a "one film for 2 years" or "one film for multiple years" system, the frequency of plastic film use will be reduced by 50% or more. This is not an approach to eliminate residual film pollution at the source but will effectively alleviate the accumulation of plastic in soil. Although the concept of "one film for 2 years" or "one film for multiple years" has been proposed in previous researches (He et al., 2009; Yan et al., 2014), its potential influences on grain yield, water use efficiency and cost-benefit balance have not been systematically explored in dryland spring maize production.

Spring maize is one of the most important grain crops in the drylands of China. While climate conditions for dryland spring maize are usually characterized by strong evaporation and rare precipitation in fallow periods, they may limit the soil moisture conditions at sowing and even lead to yield failures (Wu et al., 2017; Cai et al., 2015). Wu et al. (2017) proposed a whole season plastic mulching model to solve this problem and suggested that mulching practices during the fallow period relieved drought during the early stage of spring maize. A "one film for 2 years" or "one film for multiple years" system may have similar effects because plastic film is not removed during the fallow period. However, this effect is still not fully understood.

The proportion of humans living in poverty is extremely high on global drylands (UNDP, 2006). In China, more than 80% of absolute poverty is distributed on drylands (National Bureau of Statistics of China, 2014), and most impoverished communities depend on farmland for survival. Plastic mulching is a common agricultural practice on drylands, and improving its economic profitability would help to alleviate poverty. Hence, it is necessary to evaluate the cost-benefit balance of film use frequency based on its influence on crop yields. In fact, a "one film for 2 years" or "one film for multiple years" system will save input costs of plastic film and field management; however, its influence on crop yield is still unknown and is dependent on its influence on soil water and temperature conditions (Zhou et al., 2009; Tarara, 2000).

The objective of this research was to evaluate the effects of reducing film use frequency on maize grain yields, water use efficiency and cost-benefit balance in dryland spring maize crop production. We hypothesized that reducing film use frequency is beneficial for improving soil moisture conditions and economic benefits. To verify this assumption, we designed a "one film for 2 years" system for dryland spring maize production on the Loess Plateau, and its performance was compared with an annual film replacement treatment and a no mulch treatment.

2.3 Materials and methods

2.3.1 *Research site*

The field experiment was conducted in year 2015 and 2016 at the Shouyang rain-fed agricultural experimental station (37°45′N, 113°12′E, 1080 m altitude), Shanxi, China. The climate at the research site is semi-arid according to the UNEP classification system (UNEP, 1992). Under average climatic conditions, the area receives 480 mm of precipitation annually, about 70% of which occurs in the summer from June until September. The conventional cropping system is continuous maize cultivation. Usually, maize is sown from late April-early May and harvested in late September-early October. The soil texture is classified as loam under the USDA soil texture classification system and is classified as Calcaric Cambisol according to the world reference base for soil resources (FAO, 2006). The top 20 cm of soil had a pH of 7.8, soil organic matter content of 18.03 $g \cdot kg^{-1}$, total N of 0.85 $g \cdot kg^{-1}$, total P of 0.63 $g \cdot kg^{-1}$, and total K of 19.39 $g \cdot kg^{-1}$.

The solar radiation, rainfall amount, air temperature, relative humidity, and wind speed were obtained every half-hour using an automatic weather station (Campbell Scientific Inc., Logan, UT, USA) near the experimental plots. Solar radiation was measured with a Silicon Pyranometer (LI200X, LI-COR, Inc., Lincoln, NE, USA). Precipitation was registered with a pluviometer (RGB1, Campbell Scientific Inc., Logan, UT, USA). Air temperature and relative humidity were measured using a Vaisala probe (HMP45C, Vaisala Inc., Tucson, AZ, USA). Wind speed was measured using a cup anemometer (03002-L, R.M. Young Inc., Traverse, MI, USA). These measurements were taken 2 m above the surface of grassland and recorded in a data-logger (CR10RX, Campbell Scientific Inc., Logan, UT, USA). With those obtained variables, the reference crop evapotranspiration (ET_0, $mm \cdot d^{-1}$) was computed using the Penman-Monteith combination equation using relevant meteorological data (Allen et al., 1998).

Following the direction of FAO-56 (Allen et al., 1998), the potential evaporation during the fallow period was approximatively calculated as a product of ET_0 and the crop coefficient in the initial stage of the growing season (Kc_{ini}). In this research, Kc_{ini} was obtained graphically from Allen et al. (1998) according to the average interval between wetting events, the evaporation power ET_0, and the importance of the wetting event, and it was set as 0.4.

During the growing season of 2015, the cumulative precipitation reached 337.4 mm, which was 20% lower than the 30-year average precipitation of 421 mm in the growing season (May to September). Two peaks of precipitation occurred on the 86th day after sowing (August 3) and the 123rd day (September 8). During the 2016 corn-growing season, the cumulative precipitation was 406.1 mm, which was slightly lower (3.5%) than the average precipitation. A precipitation peak occurred on the 77th day after sowing (July 20), and it was 130.7 mm.

2.3.2 Experimental design and field management

We applied three treatments: ①Field without plastic mulching (NM) —in this system, the field was not covered by plastic film; ② Replacing film annually (PM) —the soil was partially covered by plastic film, which was replaced yearly; and ③ "One film for 2 years" (PM2) —when the crop was harvested, the plastic film was kept in place and soil tillage was not carried out. In this study, partial plastic mulchingmethod and plastic film with thickness of 10 μm was used. On the two sides of each mulched stripe band (80 cm width), no-mulched stripe bands with width of 40 cm were set to provide space for tractor to travel and farmer to walk and avoid film damage caused by wheels rolling or farmer trampling during weeding and harvest. The plastic film was therefore re-used in the second year. The experiment employed a completely randomized design with three replicates, and each plot area measured 60 m^2 (6 m× 10 m). Corresponding operation methods for NM, PM and PM2 are described below:

2.3.2.1 Tillage

Rotary tillage was carried out for all three treatments in the first year with a small walking tractor with a tillage depth of about 30 cm; for the second year, no-tillage was applied in PM2, and rotary tillage was carried out for NM and PM. Tillage was carried out about 5-10 days before sowing.

2.3.2.2 Fertilization

In accordance with local practice, fertilizers were applied at rates of N 225 kg·ha^{-1} (Urea), P$_2$O$_5$ 162 kg·ha^{-1} (Calcium superphosphate), and K$_2$O 45 kg·ha^{-1} (Potassium chloride) before sowing without topdressing, and in 2015, fertilizers were applied into the furrows in bare strips and mulch strips in PM and PM2, before the film was laid out. In 2016, to protect the plastic film in PM2, fertilizers were only applied in the furrows in bare strips.

2.3.2.3 Plastic film application and maize sowing

Clear and impermeable polyethylene (PE) film with a width of 80 cm and thickness of 10 μm was used. Two shallow furrows were dug with a spade, and then the film edges were fixed in the furrow with the excavated soil. This led to soil coverage of about 67% with the plastic film. Spring maize was sown with row distance of 60 cm and plant spacing of 30 cm (sowing density 55,556 plant·ha^{-1}). The maize cultivar "Qiangsheng 51" was sown on May 1st, 2015, and on May 5th, 2016. In filed with plastic mulching, because the plastic film was impermeable, rainfall may infiltrate into soil through three pathways: (1) part of rainfall was intercepted by the maize leaves and transferred along the stem into planting hole; for the other part, it (2) reached the ground directly and infiltrated into the bare soil, or (3) reached the surface of plastic film and horizontally flowed into the bare soil through the film side.

2.3.2.4 Seedling thinning and weeding

After seedling emergence, seedling thinning was carried out manually, and an herbicide mixture of 2,4-D butylate, paraquat, Dijie® and Baoguan® was used to control weeds.

2.3.2.5 Harvest, straw and film removal

In 2015, maize was harvested on September 30th, and in 2016, maize was harvested on October 1st. After harvest, all of the maize stubble was removed from the field manually and then used as animal fodder. In the PM treatment, the film was removed manually on October 1st in 2015. In PM2, the film was kept on the soil surface after harvest in 2015 and removed on October 2nd, 2016.

2.3.3 *Soil water content*

The soil water content was determined gravimetrically (w/w). To understand the water storage change, before sowing and after harvest in each growing season, the soil water content was determined to a depth of 2 m at 0.1-m intervals using a 0.06-m diameter hand auger in bare soil and mulched soil. Furthermore, we determined the soil water content during the growing season to a depth of 1 m (0.1-m intervals) every 10 days in order to obtain information about the soil water dynamics. This was done using a finer hand auger (0.03-m diameter), to limit soil disturbance caused by sampling as much as possible. When the weather did not allow sampling on the planned date (due to, e.g., heavy rainfall), sampling was postponed for 1–2 days. In order to take into account inherent soil heterogeneity, we randomly sampled three positions on each plot every time, and their average value was used for the final statistical analysis. After that, the volumetric water content (VWC) was obtained by multiplying the gravimetric water content with the bulk density and then divided by the water density. Soil water storage (W_s), evapotranspiration (ET), and water use efficiency (WUE) were calculated according to Cai et al. (2015):

$$W_s = \sum_i^n (h_i \times \theta_{vi}) \times 10 \tag{2-1}$$

$$ET = P - (W_{s-sowing} - W_{s-harvest}) \tag{2-2}$$

where W_s (mm) is the soil water storage for 0–200 cm; h (cm) is the depth interval of the soil sample; θ_{vi} is the soil gravimetric water content (%), i is the soil layer; n is the number of soil layers; ET (mm) is the evaporation of water from the soil surface plus transpiration from the crop; and $W_{s-sowing}$ (mm) and $W_{s-harvest}$ (mm) are the soil water storage before sowing and after harvest, respectively.

WUE (kg·ha^{-1}·mm^{-1}) was calculated as the grain yield divided by the seasonal ET.

2.3.4 *Soil temperature and soil thermal properties*

Temperature sensors (HIOKI 3633-20, Hioki E. E. Corporation, Japan) were installed in each plot at 5-cm depth between the plant rows. The soil temperature was recorded every half hour automatically from sowing until harvest, and then the mean daily temperature was calculated. The soil thermal time (TT_{soil}, ℃) was calculated using the following equation (McMaster and Wilhelm, 1997):

$$TT_{soil} = \Sigma(T_{mean} - T_{base}) \qquad (2-3)$$

where T_{base} is the base temperature of 10 ℃ for maize growth (Miedema, 1982), and T_{mean} is the daily mean soil temperature. When $T_{mean} < T_{base}$, TT_{soil} was considered to be 0 ℃, which means that this day makes no contribution to the cumulative soil thermal time.

2.3.5 Dry matter accumulation and maize yield

2.3.5.1 Above ground dry matter

For each plot composed of 12 rows, 3 plants were selected every month from the 3rd, 4th, 9th and 10th rows. The shoots were cut at ground level, and then the total shoot biomass was determined gravimetrically after oven drying at 105 ℃ for 30 min initially and then at 75 ℃ for 48 h.

2.3.5.2 Maize yield

The center 4 rows in each plot were selected to measure maize yield, and for each row, the center 5 m was manually hand-harvested in early October. The grains were manually removed from the cobs and weighed; subsamples of approximately 1 kg per plot were weighed fresh, oven-dried to a constant weight at 70 ℃ and re-weighed to determine the water content (Cakir, 2004). The grain yield per plot was also calculated on a "wet-mass basis" (standard water content of 15.5%) (Payero et al., 2008).

2.3.6 Cost-benefit analysis

According to the farmers' treatments for plastic residues, the life cycle of the plastic film can be organized in two big categories following the fate of the plastic: ① "from cradle to grave". In this category plastic residues are either not collected and are left in field (Scenario 1, Figure 2-1), or collected but burnt (Scenario 2); ② "from cradle to cradle". This type means the plastic residues are harvested by the farmer and then sold to manufacturers to recycle and produce raw material again (Scenario 3).

Based on the life cycle of the plastic film, the cost-benefit analysis can be divided into three parts: ①Manufacturing and market circulation of plastic film. This part starts from the buying of raw material until selling the plastic film to the farmers, and manufacturers and sellers are the beneficiaries of this part; ②Use by farmer and the collection. This part starts from the buying of plastic film, and ends until the crop harvest for scenario 1, while for scenario 2 and 3, this part ends until the collection of plastic residues. Farmers are beneficiaries of this part; ③Manufacturing with recycled plastic film. This part starts from the buying of plastic film from

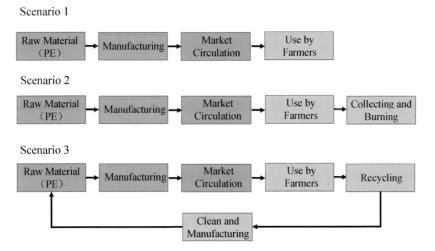

Figure 2-1 Three scenarios for the life cycle of plastic film

farmers, and ends until the productions which was produced with the recycled plastic film. The manufacturing and market circulation of plastic film and plastic residues (i.e. the first and third part) are largely influenced by many uncertain factors such as the scale of production, the salary of workers and sellers, transport cost etc. Because of the lack of data to make a complete evaluation for the first and third part, we only make the cost-benefit analysis for the second part. In other words, we focus only on the benefits of farmers from plastic mulching technology in three scenarios.

Our cost-benefit analysis adopted the cost-benefit accounting system for agricultural productions of National Development and Reform Commission (NDRC) of the government of the People's Republic of China (NDRC, 2016), in which the cost included input materials, cost for service, labor cost, and opportunity cost for self-supporting farmland or land rent, and the benefits came from agricultural productions (for scenario 1 and scenario 2), and selling of plastic residues (for senario 3). The opportunity cost for self-supporting farmland meant the lost earning from renting out farmland when farmers managed their farmland by themselves. In our case, the cost for input materials mainly included seed, fertilizer, pesticides, and plastic film; the cost for machinery service mainly included tillage, sowing, film laying and maize harvest; the cost for labour mainly included seedling thinning, herbicide spraying, straw remove, plastic film remove, grain drying and threshing.

In this study, the farm gate prices for seed, pesticides were obtained from five stores in Shouyang County. Prices for fertilizer, film, labour, maize, and opportunity cost for self-supporting farmland were obtained from government statistics in 2015 (NDRC, 2016). Prices for the collected plastic film was based on the average price offered by purchasers in 2016 (http://baojia.feijiu.net/) (Table 2-1).

It should be noted that although tillage, sowing, film laying and harvesting were completed manually in the experiment, we used the local market price of machinery service in the calculations to reflect actual production. In actual production, the film laying and maize sowing were usually completed by an integrative machine which was pulled by small wheeled tractor and could complete maize sowing and film laying at the same time, and rotary tillage was usually completed by a medium sized wheel tractor before sowing, and the harvest was usually completed by a two-line or multi-line backpack harvester. In China, farmer usually rent machinery to do those work and they pay to the machine owners by the area. The prices of machinery services were obtained through a survey in three villages of Shouyang County (Table 2-1).

Table 2-1 Prevailing prices for inputs and outputs used for calculation of cost-benefit balance

		Item	Unit	Price (US$)[1]	Data sources
Inputs	Materials	Urea	kg N	0.63	NDRC 2016
		Calcium superphosphate	kg P_2O_5	0.87	NDRC 2016
		Potassium chloride	kg K_2O	0.87	NDRC 2016
		Seed	kg	5.47	Survey
		Herbicide	l	15.63	Survey
		Plastic film	kg	1.97	NDRC 2016
	Labor	Seedling thinning	Day[3]	12.19	NDRC 2016
		Herbicide spraying			
		Straw remove			
		Film remove			
		Drying and threshing			
	Machine operation	Tillage	ha	187.50	Survey
		Sowing	ha	70.31	Survey
		Film laying	ha	46.87	Survey
		Maize harvest	ha	140.63	Survey
Opportunity cost for self-supporting farmland[2]			ha	492.19	NDRC 2016
Outputs		Maize grain	kg	0.29	NDRC 2016
		Plastic residues	kg	0.70	Survey

Note: 1. US$ 1 = 6.4 RMB Yuan, according to the average exchange rate in 2015 and 2016, Bank of China; 2. Opportunity cost for self-supporting farmland meant the lost earning from renting out farmland when farmers manage their farmland by themselves; 3. 1 Day = 8 hours for a medium labour.

Similar to Guto et al. (2011), the labour used for seedling thinning, straw remove, film remove, herbicide spraying, grain drying and threshing in this study was monitored on the trial field and corroborating them against estimates of 20 farmers neighbouring the trial site. The work rates for seedling thinning, straw remove, film remove, herbicide spraying and grain drying and threshing were estimated as 5, 15, 10, 2.5, 45 labor·day·ha^{-1}.

Used plastic film per area (Q_{film}, kg·ha^{-1}) was calculated as:

$$Q_{film} = F_{mulch} \times Thick_{film} \times \rho_{film} \times 10\ 000 \tag{2-4}$$

where F_{mulch} was the fraction of ground covered by plastic film (-); $Thick_{film}$ was the thick of plastic film (mm), ρ_{film} was the density of polyethylene (0.93 t·m^{-3}). In this study, F_{mulch} was 0.68, $Thick_{film}$ was 0.01 mm. Thus, value of Q_{film} was 63.0 kg·ha^{-1} in this study.

2.3.7 Statistical analysis

We used a one-way ANOVA to conduct analyses of variance with SAS v8.0 software (SAS Institute, Cary, NC, USA). Least significant differences (LSD) were used to detect the mean differences between the treatments.

2.4 Results

2.4.1 Effect of the "one film for two years" system on the microclimate

2.4.1.1 Soil moisture

Figure 2-2 shows the volumetric water content (VWC) at sowing during the second growing season for NM, PM, and PM2. Compared to NM and PM, the PM2 treatment effectively improved the soil moisture in the 0-10 cm ($P<0.01$) and 10-20 cm ($P<0.05$) depths. No significant difference was found for NM, PM and PM2 in the other soil layers. Furthermore, we found that both PM and PM2 had no significant influence on the soil water storage in the 0-200 cm depth.

Figure 2-3 shows the soil water dynamic under NM, PM and PM2 treatments during the two growing seasons (year 2015 and 2016). Because there was no difference in management practices between the PM and PM2 treatments during the first growing season (year 2015), their average values were compared with NM. During the first growing season, the average VWC during the entire growing season of 2015 was 21.4% in the 0-20 cm layer, 19.1% in the 20-40 cm layer and 18.6% in the 60-100 cm layer of the PM treatment. This was 0.6 percentage points (pp), 0.9 pp and 0.7 pp higher than the corresponding value in the NM treatment ($P<0.05$, $P<0.05$, and $P<0.05$). The VWC was significantly higher in PM than in NM on the 62nd and 103rd day after sowing ($P<0.01$) in the 0-20 cm and 20-60 cm layers and only on the 62nd day in the 60-100 cm layer. No significant differences were found for other sampling times and soil layers.

During the growing season of 2016, we found that the VWC in PM2 was 2.3 pp, 1.0 pp and 0.7 pp higher than under NM in the 0-20 cm, 20-60 cm and 60-100 cm depths ($P<0.01$, $P<0.01$, and $P<0.05$), and 0.8 pp, 0.5 pp, and 0.3 pp higher than PM ($P>0.05$, $P>0.05$, and $P>0.05$), respectively. On 67% of the sampled dates, there was a significant difference between PM2 and NM in the 0-20 cm depth. In the 20-60 cm layer, only

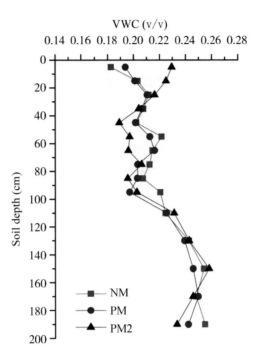

Figure 2-2 Volumetric water content (VWC) at sowing time during the second growing season (2016) on fields without plastic mulching (NM), replacing film annually (PM) and one film for two seasons (PM2) treatments

20% of the sample dates exhibited a significant difference. A significant difference between PM2 and PM was observed on the 6th day after sowing. This indicated that, compared to PM, PM2 was helpful in improving the soil moisture at the seedling stage but had little influence on the average soil moisture during the growing season.

2.4.1.2 Soil temperature

The cumulative soil thermal time (TT_{soil}) was 1,369 and 1,499 ℃ for NM and PM, respectively, for the whole growing season of 2015, and it was 1,469, 1,639 and 1,609 ℃ for NM, PM and PM2 in 2016 (Figure 2-4). Compared to NM, PM resulted in a cumulative TT_{soil} increase of 130 ℃ and 169 ℃ ($P<0.01$) in year 2015 and 2016, respectively. The difference between PM and PM2 on cumulative temperature was not significant.

Figure 2-4 also shows the evolution of TT_{soil} over the growing season in the different treatments. During the growing season, the gap of TT_{soil} between PM and NM was large in the early stage, and then it became smaller as time went on in both 2015 and 2016 as results of increasing leaves-shading with crop growth. In 2015, before the 90th day (the time for reaching maximum canopy coverage), the average daily TT_{soil} was 1.1 ℃ higher in PM than in NM ($P<0.01$), and after the 90th day, the average daily TT_{soil} was only 0.5 ℃ higher in PM ($P>0.05$). In 2016, the average daily TT_{soil} was 1.7 ℃ higher in PM than in NM ($P<0.01$)

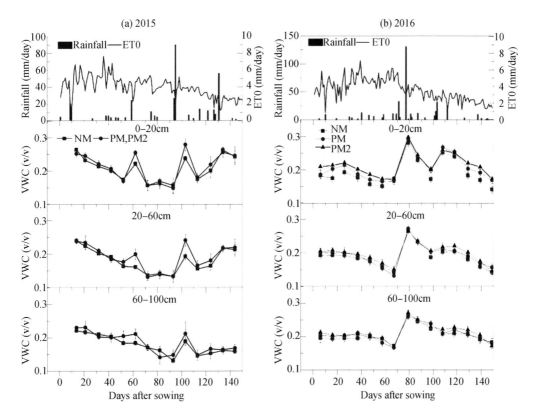

Figure 2-3 The volumetric water content (VWC) during the growing seasons of 2015 and 2016 on fields without plastic mulching (NM), replacing film annually (PM) and one film for two years (PM2) treatments

before the 90th day and only 0.3 ℃ higher after the 90th day ($P>0.05$). In PM2, the average daily TT_{soil} was 0.02 ℃ higher before the 90th day ($P>0.05$) and 0.5 ℃ lower after the 90th day ($P>0.05$) compared with PM; furthermore, TT_{soil} was 1.7 ℃ higher before the 90th day ($P<0.01$) and 0.2 ℃ lower after 90th day ($P>0.05$) compared with NM.

2.4.2 Effect of "one film for 2 years" on maize yield and water use efficiency

2.4.2.1 Dry matter accumulation

Figure 2-5 shows that the accumulation of aboveground dry matter was much quicker in mulched treatments than in NM. At the end of the growing season of 2015, the aboveground dry matter was 16% higher in PM than in NM ($P<0.01$), and in 2016, it was 12% higher in PM than in NM ($P<0.05$). No significant difference was found between PM2 and PM, and the aboveground dry matter in PM2 was 10% higher than in NM at the end of the growing season of 2016 ($P<0.05$).

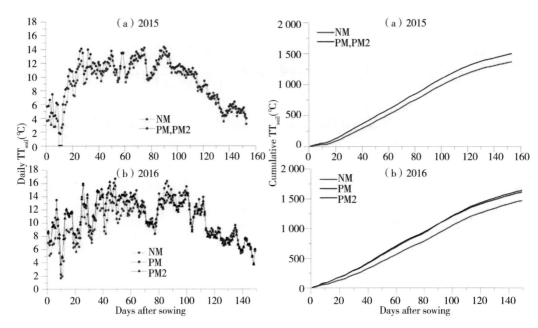

Figure 2-4 Daily and cumulative soil thermal time (TT_{soil}) during the growing season of 2015 and 2016 on fields without plastic mulching (NM), replacing film annually (PM) and one film for two seasons (PM2) treatments

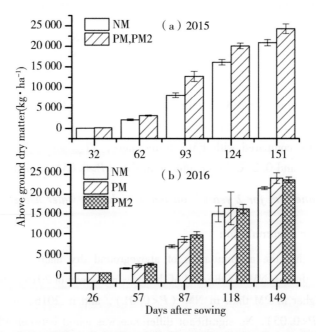

Figure 2-5 Accumulation of aboveground dry matter during the growing season of 2015 and 2016 for the field without plastic mulching (NM), replacing film annually (PM) and one film for two seasons (PM2) treatments

2.4.2.2 Maize yield, evapotranspiration and water use efficiency

Compared to PM, PM2 has no significant influence on grain yield (Table 2-2). Compared with NM, we found that PM significantly improved the grain yield by 12.1% and 25.0% in year 2015 and 2016, respectively. In 2016, the grain yield improved by 20.2% in PM2 compared to NM. For the total grain yield in 2015 and 2016, no difference between PM and PM2 was found; however, compared to NM, the total grain yield in PM2 improved by 16.3%.

The difference of evapotranspiration between PM and PM2 was not significant. Compared to NM, the WUE improved by 15.4% and 25.9% with PM in year 2015 and 2016, and it improved by 16.4% with PM2 in 2016 and by 15.5% on average in year 2015 and 2016.

Table 2-2 Maize grain yield, evapotranspiration (ET) and water use efficiency (WUE) of fields without plastic mulching (NM), replacing film annually (PM) and one film for two seasons (PM2) treatments

		NM	PM	PM2
Grain ($kg \cdot ha^{-1}$)	2015	9 464b	10 608a	10 608a
	2016	10 020b	12 529a	12 047a
	Total	19 484b	23 137a	22 656a
ET (mm)	2015	377a	368a	368a
	2016	440a	437a	454a
	Total	816a	804a	822a
WUE ($kg \cdot ha^{-1} \cdot mm^{-1}$)	2015	25.1b	29.0a	29.0a
	2016	22.8b	28.7a	26.5a
	Average	23.9b	28.8a	27.6a

Note: Numbers in each column followed by different letters indicate significant ($P \leq 0.05$) differences between treatments according to LSD tests.

2.4.3 Cost-benefit analysis

Table 2-3 shows the cost-benefit analysis of PM, PM2 and NM. Compared to PM, the PM2 treatment reduced the cost of plastic film and tillage but did not significantly reduce the benefit from maize grain, and it generated 12% and 81% higher economic profit than PM and NM in scenario 1, and 21% and 70% higher than PM and NM for scenario 2, and 17% and 74% higher than PM and NM in scenario 3 respectively. At the same time, compared to NM, profit improved by 62%, 41%, 48% in PM in scenario 1, 2 and 3 respectively for higher yields.

Table 2-3 Total cost, benefits and net benefits in 2015 and 2016 for fields without plastic mulching (NM), replacing film annually (PM) and one film for two seasons (PM2) treatments (Monetary unit: US$·ha^{-1}·2 years^{-1})

			NM	PM	PM2
Cost	Materials	Fertilizer	643	643	643
		Seed	328	328	328
		Herbicide	94	94	94
		Plastic film	0	248	124
	Labor	Seedling thinning	122	122	122
		Herbicide spraying	61	61	61
		Straw remove	366	366	366
		Film remove			
		−Scenario 1	0	0	0
		−Scenario 2	0	244	122
		−Scenario 3	0	244	122
		Drying and Threshing	1,097	1,097	1,097
	Machinery services	Tillage	375	375	188
		Sowing	141	141	141
		Film laying	0	94	47
		Maize harvest	281	281	281
	Opportunity cost for self-supporting farmland		984	984	984
Benefits	Maize grain		5,650	6,710	6,570
	Recycling of plastic film				
		−Scenario 1	0	0	0
		−Scenario 2	0	0	0
		−Scenario 3	0	88	44
Net benefits	Scenario 1		1,159	1,876	2,096
	Scenario 2		1,159	1,632	1,974
	Scenario 3		1,159	1,720	2,018

Note: In scenario 1, plastic film is not collected; in scenario 2, plastic film is collected but not sold out; in scenario 3, platic film is collected and sold out. Those three scenario are three possible ways to treat the plastic residues in actual production.

2.5 Discussion

During the growing season of year 2015 and 2016, we observed relative higher soil water content in plastic mulching treatment than NM. This was in accordance with previous research (Gong et al., 2017; Zhou et al., 2009) which suggested that plastic mulching was effective at reducing soil evaporation and then improving the soil water content during the growing season. Our results indicated that PM2 improved the soil moisture at sowing time and at the seedling stage in the 0–20 cm layer compared to PM. For most farmland in northern China, during the fallow period of spring maize (from October to April of the next year), the field water balance was usually negative because of limited precipitation (Guo et al., 2012; Piao et al., 2010). At our research site, the calculated potential evaporation was 167 mm during the fallow period between 2015 and 2016, and the observed precipitation was 121 mm. This result was consistent with previous research by Wu et al. (2017) and Cai et al. (2015) which suggested that mulching practices during the fallow season could relieve drought stress that occurs at sowing time and in the earlier stages of maize growth. However, we did not observe significantly different ET between PM, PM2 and NM for the overall growing season. Gong et al. (2017) reported that ET decreased by 9.3% under plastic mulching on the Loess plateau, whereas Fan et al. (2017) and Zhang et al. (2011) found that ET was not significantly reduced from plastic mulching and even increased. Those studies suggest that the effects of plastic mulching on ET may be influenced by environment variables. In PM2, although some holes appeared and the film was partly destroyed by weeds during the early stage of the growing season (Figure 2-6), no significant difference in seasonal average soil moisture was observed between PM and PM2. This phenomenon could be explained by three reasons: ①the extent of film degree was not enough to induce a significant drop in soil moisture. The holes were only a very small part of the whole film, and most soil was still covered by plastic film; ② the improvement in soil water content at planting may offset the soil moisture drop caused by film damage; ③the effect of plastic mulching was only prominent at the early stage of the growing season (Zhou et al., 2009; Li et al., 2012), which meant that the further film damage in the late growing season was negligible.

Figure 2-6 Film destruction in the "one film for two seasons" treatment (b) during the early stage of the second growing season and a comparison with the replacing film annually treatment (a)

At the same time, we also did not observe significant difference between PM and PM2 on soil temperature. It was true that the light transmittance of transparent PE decreases over time because of dust accumulation and aging (Castellano et al., 2008); however, PM2 reduced the accumulation of drops beneath the film because of the existence of small holes (Figure 2-

6). Moreover, because of the growth of weeds below the film in PM2, it established a larger insulating air gap, and greater heat storage, or less heat loss, may have occurred (Ham et al., 1993; Ham and Kluitenberg, 1994). Furthermore, due to the effects of evaporation, PM2 still could improve the soil temperature because of the reduction in latent heat flux (Liu et al., 2010). Further quantitative research is needed to reveal the soil water and heat flux and their loop in PM2 and to explain the interaction between different factors. Furthermore, our results confirmed that PM2 has equivalent performance to PM in terms of soil moisture and temperature adjustment.

Previous research indicated that modifications in plastic mulching on the microclimate were able to reduce the temperature and water stress and then led to improvements in crop yield (Qin et al., 2015; Fan et al., 2017; Zhang et al., 2011), and this was confirmed by our research. Moreover, no significant difference in maize yield was found from PM and PM2. Although the soil water content improved from PM2 in 0–20 cm before sowing and at the seedling stage, it seemed to have little influence on the average soil moisture over the whole growing season and on the ET in our research site. Soil moisture and temperature were the two most important factors influencing crop growth (Raes et al., 2009), and similar soil moisture and temperature dynamics in PM and PM2 may be able to explain their consistency in maize yield. This result was similar to Wu et al. (2017) in which the advantage of mulching throughout the whole season on maize yields was only found in one of three tested years.

Our results indicated that the WUE was significantly improved by PM and PM2 and that there was no significant difference between PM and PM2. This finding agreed with previous research (Liu et al., 2010; Xu et al., 2015; Qin et al., 2015). Liu et al. (2010) reported that the WUE of maize improved by 23–25% in a two-year experiment on the Loess Plateau. Xu et al. (2015) reported that the WUE of maize increased by 16% in plastic mulching treatment at five sites in northeastern China. Qin et al. (2015) reported that the mean effect of plastic mulching on WUE was 81% at high N input and 30% at low N input for maize in a meta-analysis. With similar ET in PM, PM2 and NM, the impermeable barrier in PM and PM2 was probably effective in reducing the evaporation and increasing the physiologically significant canopy transpiration and plant productivity (Liu et al., 2010). Moreover, by applying PM2, the significant improvement in soil moisture in 0–20 cm meant that PM2 truly reduced water evaporation during the fallow period and made water resources more available for crop growth in the fallow period. The improvement in topsoil moisture was especially important for the seedling stage because the maize roots were mainly distributed in the topsoil during the seedling stage (Chassot et al., 2001). However, in the second year of this research, although the soil water content in the 0–20 cm layer was lower in PM, it immediately improved from the rainfall, and severe spring drought did not occur. This may explain why PM and PM2 had similar WUE in this study. However, considering that the frequency of agricultural drought is increasing (Leng et al., 2015; He et al., 2016; Piao et al., 2010), the effects of PM2 on WUE were probably more predominant in drought years.

Our research suggests that PM2 was effective at improving economic profits. In fact, canceling maize price protection in China led to reductions in maize prices, and cost savings practices became more important than yield improvements for profit generation. Compared to PM, the cost for plastic film, tillage, film laying, film remove were reduced in PM2. This was the main reason for the improvement in profits in PM2. Prices and difference of maize yield between PM and PM2 may vary with time and regions. Thus, we carried out a sensitivity analysis for the response of advantage of PM2 over PM on net benefits (i.e. net benefits in PM2 minus net benefits in PM) to the change of different prices (Table 2-4). Corresponding results can be used to compare the benefits of PM and PM2 in places where the prices are different from the selected site in this study.

Table 2-4 Response of PM2 over PM on net benefits (i.e. net benefits in PM2 minus net benefits in PM) to per 10% change of prices

	Scenario 1	Scenario 2	Scenario 3
Price of plastic film	5.7%	3.6%	4.2%
Labor price	—	3.6%	4.1%
Tillage price	8.6%	5.5%	6.3%
Film deposit price	2.2%	1.4%	1.6%
Maize grain price	-6.4%	-4.1%	-4.7%
Price of collected film residues	—	—	-1.5%

Note: In scenario 1, plastic film is not collected; in scenario 2, plastic film is collected but not sold out; in scenario 3, platic film is collected and sold out. Those three scenario are three possible ways to treat the plastic residues in actual production.

The shortcoming of this research was that we just tested the performance of "one film for 2 years". In fact, on the basis of current research, "one film for multiple years" may have more advantages for plastics pollution control and economic benefits. However, the "one film for multiple years" system called for an innovative design of plastic film that has good durability, weathering ability and high tensile strength (to reduce film destruction caused by wind and weeds). In fact, to control "white pollution", some local governments in China (such as Xinjiang, Gansu, and Ningxia province) have released new mandatory standards for plastic film, and films with a thickness less than 0.010 mm have stopped being used in those places and the government encouraged farmers to use film with good durability and weathering ability. This provided favorable conditions for the application of a "one film for 2 years" or "one film for multiple years" system. On the other hand, because of cost savings in the "one film for 2 years" or "one film for multiple years" system, promotion of a new type of film without an added burden to farmers became possible. However, the design of film for a "one film for multiple years" system and evaluation of its agricultural and ecological effects require further research.

2.6 Conclusion

In this study, the influences of "one film for 2 years" system on soil moisture, temperature, maize yield, water use efficiency and cost-benefit balance were evaluated on the Loess Plateau. The results suggested: Compared to PM, PM2 significantly improved the soil moisture in the 0-20 cm layer at planting and at the seedling stage, and this effect did not induce an increase in the average soil moisture and ET for the overall growing season; PM2 had no significant influence on the cumulative soil temperature compared to PM, however, compared to NM, the cumulative soil temperature improved by 140 ℃; PM2 had no significant influence on maize yield and WUE compared to PM, and compared to NM, they improved by 16.3% and 15.5%, respectively; Because of the lower cost of plastic film and tillage, and due to similar maize yields to PM, PM2 leads to 12% and 81% higher economic profit as compared to PM and NM in scenario1 (without film collection), and 21% and 70% higher in scenario 2 (with film collection but without selling the plastic residues), and 17% and 74% higher in scenario 3 (with film collection and plastic residues selling), respectively.

2.7 Reference

Allen R. G., Pereira L. S., Raes D., et al, 1998. Crop evapotranspiration: guidelines for computing crop water requirements. FAO Irrigation and Drainage [M]. Paper no. 56. Rome: Food and Agriculture Organization of the United Nations.

Cai T., Zhang C., Huang Y., et al, 2015. Effects of different straw mulch modes on soil water storage and water use efficiency of spring maize (Zea mays L.) in the Loess Plateau of China [J]. Plant Soil Environ., 61 (6): 253-259.

Cakir R, 2004. Effect of water stress at different development stages on vegetative and reproductive growth of corn [J]. Field Crop Res., 89 (1): 1-16.

Castellano S., Mugnozza G. S., Russo G., et al, 2008. Plastic nets in agriculture: a general review of types and applications [J]. Appl. Eng. Agric., 24 (6): 799-808.

Chassot A., Stamp P., and Richner W, 2001. Root distribution and morphology of maize seedlings as affected by tillage and fertilizer placement [J]. Plant Soil, 231 (1): 123-135.

Chen Y. S., Wu C. F., Zhang H. B., et al, 2013. Empirical estimation of pollution load and contamination levels of phthalate esters in agricultural soils from plastic film mulching in China [J]. Environ. Earth Sci., 70 (1): 239-247.

Dong H., Liu T., Li Y., et al, 2013. Effects of plastic film residue on cotton yield and soil physical and chemical properties in Xinjiang [J]. Trans. Chin. Soc. Agric. Eng., 29 (8): 91-99. Chinese.

Fan Y. Q., Ding R. S., Kang S. Z., et al, 2017. Plastic mulch decreases available energy and evapotranspiration and improves yield and water use efficiency in an irrigated maize cropland [J]. Agric. Water Manage., 179 (1): 122-131.

FAO, 2006. Word reference base for soil resources 2006-a framework for international classification, correlation and communication. World Soil Resources Reports [M]. Rome: Food and Agriculture Organization.

Godfray H. C. J., Beddington J. R., Crute I. R., et al, 2010. Food security: the challenge of feeding 9 billion people [J]. Science, 327 (5967): 812-818.

Gong D., Mei X., Hao W., et al, 2017. Comparison of ET partitioning and crop coefficients between partial plastic mulched and non-mulched maize fields [J]. Agric. Water Manage., 181: 23-34.

Guo S. L., Zhu H. H., Dang T. H., et al, 2012. Winter wheat grain yield associated with precipitation distribution under long-term nitrogen fertilization in the semiarid Loess Plateau in China [J]. Geoderma, 189 (6): 442-450.

Guo Y. F., Li S. Y., and Huo Y. Z, 2016. The effects of different residual film amount on spring maize production traits and soil moisture [J]. Water Sav. Irri., 4: 47-49. Chinese.

Guto S. N., Pypers P., Vanlauwe B., et al, 2011. Tillage and vegetative barrier effects on soil conservation and short-term economic benefits in the central kenya highlands [J]. Field Crop Res., 122 (2): 85-94.

Ham J. M., Kluitenberg G. J., and Lamont W. J, 1993. Optical properties of plastic mulches affect the field temperature regime [J]. J. Am. Soc. Hortic. Sci., 118 (2): 188-193.

Ham J. M., and Kluitenberg G. J, 1994. Modeling the effect of mulch optical properties and mulch-soil contact resistance on soil heating under plastic mulch culture [J]. Agr. Forest Meteorol., 71 (3): 403-424.

He J., Yang X. H., Li Z., et al, 2016. Spatiotemporal variations of meteorological droughts in china during 1961-2014: an investigation based on multi-threshold identification [J]. Int. J. Disaster Risk Sci., 7 (1): 63-76.

He W. Q., Yan C. R., Zhao C. X., et al, 2009. Study on the pollution by plastic mulch film and its countermeasures in China [J]. J. Agro-Environ Sci., 28: 533-538. Chinese.

Leng G. Y., Tang Q. H., and Rayburg S, 2015. Climate change impacts on meteorological, agricultural and hydrological droughts in China [J]. Global Planet. Change, 126 (126): 23-34.

Li R., Hou X. Q., Jia Z. K., et al, 2012. Effects of rainfall harvesting and mulching technologies on soil water, temperature, and maize yield in Loess Plateau region of China [J]. Soil Res., 50 (2): 105-113.

Liu E. K., He W. Q., and Yan C. R, 2014. "White revolution" to "white pollution"-agricultural plastic film mulch in China [J]. Environ Res. Lett., 9 (9): 3.

Liu Y., Yang S. J., Li S. Q., et al, 2010. Growth and development of maize (Zea mays L.) in response to different field water management practices: Resource capture and use efficiency [J]. Agr. Forest Meteorol., 150 (4): 606-613.

McMaster G. S., and Wilhelm W. W, 1997. Growing degree-days: one equation, two interpretations [J]. Agr. Forest Meteorol., 87 (4): 291-300.

Miedema P, 1982. The effects of low temperature on Zea mays [J]. Advan. Agron., 35: 93-128.

National Bureau of Statistics of China (CN). 2014. China Statistical Yearbook [M]. Beijing: China Statistics Press. Chinese.

[NDRC] National Development and Reform Commission (CN). 2016. Compilation of national cost and income of agricultural products 2016 [M]. Beijing: China Statistics Press. Chinese.

Niu W., Zou X., Liu J., et al, 2016. Effects of residual plastic film mixed in soil on water infiltration, evaporation and its uncertainty analysis [J]. Trans. Chin. Soc. Agric. Eng., 32: 110-119. Chinese.

Payero J. O., Tarkalson D. D., Irmak S., et al, 2008. Effect of irrigation amounts applied with subsurface drip irrigation on corn evapotranspiration, yield, water use efficiency, and dry matter production in a semiarid climate [J]. Agric. Water Manage., 95 (8): 895-908.

Piao S., Ciais P., Huang Y., et al, 2010. The impacts of climate change on water resources and agriculture in China [J]. Nature, 467 (7311): 43-51.

Qin W., Hu C. S., and Oenema O, 2015. Soil mulching significantly enhances yields and water and nitrogen use efficiencies of maize and wheat: a meta-analysis [J]. Sci. Rep-UK., 5: 13.

Raes D., Steduto P., Hsiao T. C., et al, 2009. AquaCrop the FAO crop model to simulate yield response to water: II. Main algorithms and software description [J]. Agron. J., 101 (3): 438-447.

Tarara J. M, 2000. Microclimate modification with plastic mulch [J]. Hort Science, 35 (2): 169-180.

Tilman D., Balzer C., Hill J., et al, 2011. Global food demand and the sustainable intensification of agriculture [J]. Proc. Natl. Acad. Sci., 108 (50): 20260-20264.

UNDP, 2006. Human development report 2006. Beyond scarcity: povert, poverty and the global water crisis [M]. New York (NY): Palgrave Macmillan.

UNEP, 1992. World atlas of desertification [M]. London: Edward Arnold.

Wu Y., Huang F. Y., Jia Z. K., et al, 2017. Response of soil water, temperature, and maize (Zea may L.) production to different plastic film mulching patterns in semi-arid areas of northwest China [J]. Soil Till. Res., 166: 113-121.

Xu J., Li C. F., Liu H. T., et al, 2015. The effects of plastic film mulching on maize growth and water use in dry and rainy years in northeast China [J]. Plos One, 10 (5): e0125781.

Yan C. R., He W. Q., and Neil C, 2014. Plastic-film mulch in Chinese agriculture: importance and problems [J]. World Agric., 4 (2): 32-36.

Zhang S. L., Li P. R., Yang X. Y., et al, 2011. Effects of tillage and plastic mulch on soil water, growth and yield of spring-sown maize [J]. Soil Till. Res., 112 (1): 92-97.

Zhou L. M., Li F. M., Jin S. L., et al, 2009. How two ridges and the furrow mulched with plastic film affect soil water, soil temperature and yield of maize on the semiarid Loess Plateau of China [J]. Field Crop Res., 113 (1): 41-47.

3

Modeling of Soil Water Dynamic and Field-scale Spatial Variation of Soil Water Content

3.1 Abstract

Numerical solution of the Richards equation with Hydrus-2D model is a low cost and fast way to get information on field-scale spatio-temporal soil water dynamics. Previous research on irrigated fields with plastic mulching have used two different approaches to take into account rainfall infiltration in Hydrus-2D: either the water falling on the plastic strip was completely neglected or it was supposed to infiltrate at the border of the plastic strip. Nevertheless, the performance of these approaches has not yet been evaluated in rain-fed fields. Considering the much more dominant role of rainfall infiltration in rain-fed agriculture, we tested an additional approach which comprises a bare strip, plastic strip and planting hole to take into account the effect of the rainfall canopy redistribution and film side infiltration, and we compared its performance to the two existing approaches. Neglecting the water falling on the plastic strip completely failed to reproduce the soil water dynamics in all soil layers under the plastic strip and in the deep soil layers of bare strip. Adding only film-side infiltration overestimated the soil water content (SWC) in 0-20 cm of the bare strip, while the performance of our proposed approach with planting hole and canopy redistribution was acceptable in different positions. After that, we compared the soil water distribution between a no-mulched field and plastic mulched field with this best performing infiltration simulation. Our simulation shows that the highest SWC in a partially plastic mulched field occurs near the planting hole. The SWC in the center zone of the mulched part was lower, while the SWC in the bare strip was lowest. Results showed that, PM improves the soil water availability not only in the plastic strip but also in the bare strip as compared to an unmulched field.

Keywords: Soil water; Hydrus 2D; Canopy redistribution; Rainfall infiltration; Spatial variation.

3.2 Introduction

Rain-fed agriculture, makes up approximately 80% of global cropland and produces 60-70% of the world's food (Falkenmark and Rockström, 2004; Rost et al., 2008). It plays a dominant role in the global food supply, especially considering the increasing global water shortage (Rockström et al., 2010). In China, rainfed agriculture accounts for approximately 25 Mha of the arable land and is mainly located on the semi-arid Loess Plateau and in northeast China, where the crop yields are limited by the soil water deficit and the low soil temperatures in the spring (Deng et al., 2008; Xiao et al., 2016). Plastic mulching has became a popular agricultural technology in those area, since it is thought to maintain soil moisture and increase the soil temperature (Tarara et al., 2000; Kader et al., 2017). For economic reasons, farmers usually do not cover the entire field with plastic film, but use polyethylene (PE) plastic strips which are alternated with bare strips, a technique called partial plastic mulching.

Soil water dynamics in fields with plastic mulching has received wide attention (Fisher, 1995; Wu et al., 2017; Ren et al., 2017; Dong et al., 2009; Zhao et al., 2012), since not only crop performance, but also environmental processes such as nitrate leaching and the emission of greenhouse gasses depend heavily on soil moisture dynamics (Qin et al., 2015; Filipovic et al., 2016; Liu et al., 2016). Measurement methods such as gravimetric soil water determination on soil samples and time domain reflectometry, and simulations have been used to obtain information on soil water dynamics and distribution (Wu et al., 2017; Ren et al., 2017; Filipovič et al., 2016; Li et al., 2015; Liu et al., 2013). Compared to field measurements, modeling studies are low-cost and present a high temporal and spatial resolution.

Models that can be used to obtain soil water information mainly include WaSim-ETH, Community Land Model, SiSPAT-Isotope, Hydrus and crop models such as SWAP, CERES, WOFOST and Aquacrop (Vereecken et al., 2016). Hydrus is mainly used at the pedo up to field scale. Crop models are mainly applied at field scale or regional scale, while the WaSim-ETH, Community Land Model, SiSPAT-Isotope are mainly applied at catchment scale or landscape scale. In fileds with plastic mulching, the different water infiltration and evaporation characteristics of plastic strips and bare strips may lead to obvious inhomogeneous soil water distribution. However, the soil water transport modules in most crop models are one-dimensional, and they are un-able to describe the field-scale spatial variation of soil water. Soil hydraulic model such as Hydrus-2D model can solve this problem by applying different type of boundary conditions for the plastic strip and bare strip. For these reasons, the Hydrus-2D model was adopted for the current study.

Hydrus-2D has been widely used to simulate soil water dynamic in irrigated field with plastic mulching, but not yet under rain-fed conditions. Different from the irrigated field, rainfall isthe only source of soil water in the rain-fed field and a good understanding of rainfall infiltration processes is decisive for a correct prediction of the field-scale spatio-temporal patterns of soil water. In case of partial plastic mulching, rainfall can infiltrate into the soil through three pathways: ①interception by maize leaves and transfer along the stem into the planting hole (i.e. canopy redistribution). Rainfall that is not intercepted by the maize leaves will ②reach the ground directly and infiltrate into the bare soil, or③reach the surface of the plastic film and flow towards the bare soil and infiltrate into the soil at the film side (film-side infiltration) (Chen et al., 2018). Previous studies have applied two approaches to represent the rainfall infiltration in irrigated fields with plastic mulching: ①In arid regions (annual rainfall amount< 200 mm), Han et al. (2015), Li et al. (2015) and Liu et al. (2013) simplified reality by neglecting canopy redistribution and film side infiltration. They assumed that rainfall directly reached the bare strip and infiltrated there (without canopy redistribution) and they omitted rainfall that reached the plastic strip. ②In the humid regions (annual rainfall amount>800 mm), Filipovic et al. (2016) and Dusek et al. (2010) also neglected canopy redistribution, but integrated the process of film side infiltration by increasing the rainfall infiltration amount in bare strip with a factor which was equal to the ratio of plastic strip width to bare strip width. However, the performance of those two methods has not been evaluated under rain-fed conditions, where the redistribution may play an even bigger role, since no additional water is added under the plastic sheets through irrigation tubes.

In this study, we developed and applied a modeling strategy taking into account of both canopy redistribution and film side infiltration in field with plastic mulching. The objectives of this study were therefore ①to compare the performance of different simulation strategies for rainfall infiltration in rain-fed field with plastic mulching; ②to quantify the field-scale spatial variation of the SWC in rain-fed field with plastic mulching using the optimized simulation strategy. We hypothesized that rainfall redistribution and film side infiltration play important roles in soil water dynamics, and are not negligible under rain-fed conditions. In this study, the performance of Hydrus-2D was tested with 2 years of field data from a spring maize (*Zea may L.*) field that located on the Loess Plateau of China. We chose to focus on maize, since it was one of the main crops in the studied region and of great importance for Chinese agriculture (38.1 million ha in China in 2015, NBSC, 2016).

3.3 Materials and methods

3.3.1 *Research site*

The research site (37°45′N, 113°12′E, 1202 m altitude) is located at the Shouyang County which belongs to Shanxi Province, at the eastern part of the Loess Plateau, about 500 km west from the China's capital Beijing. The majority of the farmland in Shouyang County produces crops, especially maize, soy bean and potato. The research area is characterized by a semiarid temperate continental monsoon climate with four distinct seasons. According to the weather record from Shouyang weather station, which is located at 15 km from our research site, the experimental site has a mean annual air temperature of 7.4 ℃, a mean annual frost-free period of 140 days and a mean annual rainfall amount of 480 mm during the past 48 years (year 1967-2014). During the experimental year 2015 and 2016, the rainfall amount was 386 mm and 461 mm, which was 20% and 4% lower than the historical mean, respectively.

The soil texture is sandy loam, and the soil is classified as a calcaric Cambisol according to the world reference base for soil resources (FAO, 2006). The upper 20 cm of soil has a pH of 7.8, a soil organic matter content of 18.03 $g \cdot kg^{-1}$, total N of 0.85 $g \cdot kg^{-1}$. The soil profile has three horizons: 0-20 cm, 20-60 cm and 60-100 cm. The topography of experimental field is flat, and the groundwater is at depths deeper than 150 meters below the surface (Gong et al., 2017). For the soil hydraulic parameters, please see 3.3.5.2.

Spring maize (*Zea mays L.*) is sown every year on the research site, and typically no crop rotation is applied. Usually, maize is sown in late April or early May and harvested in early October. After harvest, there is a fallow period until sowing in the next year. The early stage of maize growing season is usually characterized by low temperature and few rainfall and accompanied by high risk of spring drought and spring chill, and high temperature and heavy rainfall mainly happen at the middle of maize growing season, while temperature and rainfall become lower at the late growing season (Gong et al., 2015). In 2015, the maize was sown in May 1st and harvested in September 30th, and in 2016, the maize was sown in May 5th and harvested in October 1st.

3.3.2 Field experiment

Two treatments were selected in this study: a no mulch system (NM) and a partial plastic mulching system (PM). Each treatment has three replications, and each replication was carried out in a plot with a width of 6 m and a length of 10 m (60 m²). As shown in Figure 3-1, the PM system included a plastic strip with a width of 80 cm and a bare strip with a width of 40 cm. PE film with thickness of 10 μm was applied in the PM treatment. For both the PM and NM treatments, the planting spacing was 30 cm, the row distance was 60 cm and the sowing density was 56,000 plants·ha^{-1}. For the detailed filed manage practices in NM and PM, please see Chen et al. (2018).

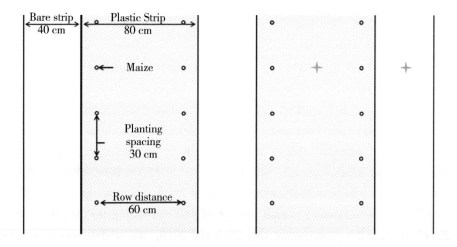

+ Soil sampling points for soil moisture measurement

Figure 3-1 Field layout of the partial plastic mulching treatment and the locations of measurement points for soil moisture

During the growing season, the soil water content was determined gravimetrically to a depth of 1 m at 0.1-m intervals at the middle of plastic strip and at the middle of bare strip. Theoretically, the higher temporal resolution the soil moisture measurement was, the more favorable for the model evaluation. However, considering the cost of gravimetric determination and the rare rainfall in semi-arid area, the soil moisture was measured every 10 days. When the weather did not allow sampling at the planned date (e. g., due to heavy rainfall), sampling was postponed for 1-2 days. As current version Hydrus-2D do not include the crop growth module, we used the measured leaf area index (LAI) to calculate the potential evaporation and transpiration together with the field-recorded weather variables. Measured LAI has been reported in our previous study, where we observed PM accelerated the development of LAI in year 2015 and 2016 (Chen et al., 2018). For the LAI measurement method and details on how we obtained weather variables (solar radiation, amount of rainfall, air temperature, relative humidity, and wind speed), please see Chen et al. (2018).

3.3.3 *Governing equations for soil water transport*

We used HYDRUS-2D (Šimůnek et al., 1994) to simulate the soil water transport. The code used the Galerkin finite element method to numerically solve the governing Richards equation (Richards, 1931):

$$\frac{\partial \theta(h)}{\partial t} = \frac{\partial}{\partial x}\left[K(h)\frac{\partial h}{\partial x}\right] + \frac{\partial}{\partial z}\left[K(h)\frac{\partial h}{\partial z} + K(h)\right] - S(h) \quad (3-1)$$

where θ was the volumetric soil water content ($cm^3 \cdot cm^{-3}$); h was the pressure head (cm); $K(h)$ was the unsaturated hydraulic conductivity ($cm \cdot day^{-1}$); t was time (day); x and z were the horizontal and vertical coordinates (cm); S was the sink term (day^{-1}).

The soil water retention curve and the unsaturated hydraulic conductivity function were estimated using van Genuchten-Mualem constitutive relationships (van Genuchten, 1980):

$$S_e(h) = \frac{\theta - \theta_r}{\theta_s - \theta_r} = \frac{1}{[1+(\alpha|h|)^n]^m} \quad (3-2)$$

$$K(h) = K_s S_e^l \left[1 - (1 - S_e^{1/m})^m\right]^2 \quad (3-3)$$

where S_e was the degree of saturation, h was the water potential (cm), θ was the volumetric water content ($cm^3 \cdot cm^{-3}$), θ_s and θ_r were the saturated and residual water contents ($cm^3 \cdot cm^{-3}$), respectively, while m (-) and n ($m=1-1/n$) (-) were shape parameters, l was a pore connectivity parameter (-), Ks was the saturated hydraulic conductivity ($cm \cdot day^{-1}$).

The root water extraction was computed according to the Feddes model (Feddes et al., 1978):

$$S(h) = \tau(h) \cdot \beta(x, z) \cdot T_p L_t \quad (3-4)$$

where T_p was the potential transpiration rate ($cm \cdot day^{-1}$), L_t was the surface length associated with transpiration (cm), $\beta(x, z)$ was the root water uptake distribution function (cm^{-2}) and $\tau(h)$ was the root water uptake stress reduction function ($0 < \tau < 1$). In Hydrus, the root water uptake was assumed to be zero when the pressure heads close to saturation ($h1$). When the pressure heads below the wilting point pressure head ($h4$), water uptake was also assumed to be zero. Water uptake was considered optimal between pressure heads

$h2$ and $h3$, whereas for pressure head between $h3$ and $h4$ (or $h1$ and $h2$), water uptake decreased (or increased) linearly with heads. Hydrus-2D provided a database of root water uptake parameters for different crops.

3.3.4 Three methods to represent rainfall infiltration in field with plastic mulching

Figure 3-2 (A) shows the two methods which were proposed by previous studies (Han et al., 2015; Li et al., 2015; Liu et al., 2013; Filipovic et al., 2016; Dusek et al., 2010): "BP" -bare strip and plastic strip; "BP+" -bare strip and plastic strip with integrating the process of film side infiltration by increasing the rainfall infiltration amount in bare strip. The impervious plastic cover was represented by a "no-flux" boundary and the bare soil strip was defined as an "atmospheric" boundary. "No-flux" boundary was applied at left and right sides with assumption of symmetry of the soil water pressure head inside and outside the geometry domain. For BP, the rainfall amount in the bare strip equaled the rainfall amount in meteorological record. For BP+, the rainfall quantity in the bare strip (R_{bare}, mm) was calculated as:

$$R_{bare} = R_{met} \times (W_{plastic} + W_{bare}) / W_{bare} \qquad (3-5)$$

where the R_{met} was the rainfall amount in meteorological record (mm), $W_{plastic}$ was the width of plastic strip (cm) and W_{bare} was the width of the bare strip (cm).

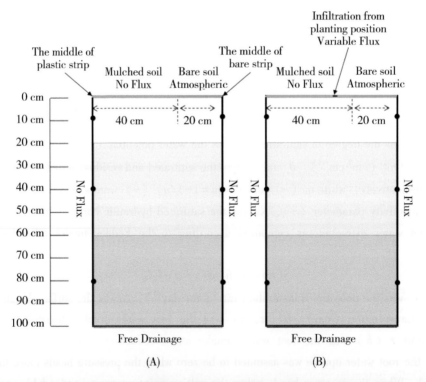

Figure 3-2 Scheme of boundary conditions in this modeling study: (A) BP and BP+; (B) BPH

In this study, we proposed a third, new strategy which took into account the rainfall canopy redistribution and film side infiltration: "BPH" —bare strip, plastic strip and planting hole method. As shown in Figure 3-2 (B), the "no-flux" boundary was imposed on the plastic strip and the "atmospheric" boundary was imposed on the bare strip. Unlike the other two methods, a 2 cm "variable flux" boundary was imposed in the planting hole position to take the rainfall redistribution caused by canopy. The canopy redistribution was defined as on the one hand the rainfall intercepted by the maize leaves and transferred along the stem (stem-flow) and on the other hand the rainfall which reached the ground directly (through-fall). In our case, we adopted the relationship of stem-flow/through-fall (S/T) and incident rainfall (R, mm) proposed by Martello et al. (2015). This relationship assumed that the S/T decreased logarithmically at increasing values of incident rainfall under a closed maize canopy and during the growth period, it increased with the maize canopy cover:

$$S/T = a \cdot [\ln(R) + b] \quad (\text{if } LAI \geqslant LAI_{closed}) \quad (3-6)$$

$$S/T = (LAI/LAI_{closed}) \cdot [a \cdot \ln(R) + b] \quad (\text{if } LAI < LAI_{closed}) \quad (3-7)$$

where S/T was the ratio of stem-flow to through-fall, LAI was the leaf area index, and LAI_{closed} was the leaf area index when the maize canopy reached the closed condition. The a and b were empirically fitted parameters, and they were -1.356 and 7.418 respectively (Martello et al., 2015).

Then, the rainfall infiltration in the bare strip (R_{bare}) was calculated by:

$$R_{bare} = R_{met} \times [1/(S/T+1)] \quad (3-8)$$

where R_{met} was the rainfall amount in meteorological record (mm), and S/T was the ratio of stem-flow to through-fall (-).

The rainfall infiltration from the planting hole (R_H) was calculated as:

$$R_H = R_{met} \times [1 - 1/(S/T+1)] \times (W_{plastic} + W_H + W_{bare})/W_H \quad (3-9)$$

where W_H was the width of planting hole (cm).

In our case, because the maize was planted near the plastic film side (Figure 3-1) and to simplify our calculation, we did not set a special boundary for film side. Instead, we assumed the film side infiltration took place in the planting hole position. Thus the rainfall infiltration from the film side (R_{FS}) was calculated as:

$$R_{FS} = R_{met} \times [1/(S/T+1)] \times W_{plastic}/W_H \quad (3-10)$$

As the results showed that the scenario BPH was the most realistic, we applied this scenario to compare PM with NM in the rest of this study. In NM, a similar 2 cm "variable flux" boundary was imposed in the sowing position to take the rainfall redistribution caused by canopy, and other parts of the upper boundary were expressed as an "atmospheric" boundary.

3.3.5 Model parameterization

3.3.5.1 Daily potential evaporation and transpiration

With the collected meteorological data, the reference crop evapotranspiration (ET_0) was computed using the Penman-Monteith combination equation (Allen et al., 1998). Then the crop evaoptranspiration was calculated by multiplying ET_0 by crop coefficient (K_c) (Allen et al., 1998). The K_c for different growing stage was determined with eddy covariance system in the same research site during 2011–2013 (Gong et al., 2017). The K_c at initial stage, middle stage, and late stage were 0.22, 0.91 and 0.94 for PM, and 0.27, 1.01 and 0.99 for NM. Potential evaporation and transpiration could be calculated separately from crop evaoptranspiration using Beer's law which partitioned the solar radiation component of the energy budget via interception by the leaf area index (Ritchie, 1972. In Hydrus-2D, calculated daily potential evaporation was applied in "atmospheric" boundary to calculate the actual evaporation, and potential transpiration was applied in Eq. 8 and Eq. 5 to calculate the actual root water uptake.

3.3.5.2 Soil hydraulic parameters

In each horizon, three un-disturbed soil samples were taken in the fallow period between the 2015 and 2016 growing seasons with brass rings 7.8 cm in diameter and 2.0 cm in height and then placed on a pressure plate apparatus following the procedures described by Dane and Hopmans (2002). The soil water content was measured at matric potentials of −10cm, −40cm, −70cm, −100cm, −300cm, −700cm, −1,000, −5,000cm, and −15,000 cm H_2O, and then, the values of parameters θ_r, θ_s, α and n were estimated by the RETC code (van Genuchten et al., 1991). Measured soil water content at different pressure head and the fitted soil water retention curves were shown in Figure 3-3. We used the fitted RETC parameters as a starting value to perform an inversion for the hydraulic parameters based on the observed soil water content in NM in 2015 which has been reported in Chen et al. (2018). The measured and inversed soil hydraulic parameters were shown in Table 3-1. In Hydrus-2D, the soil hydraulic parameters were applied in Eq. 5-7 to describe the water continuous transport in soil medium.

Table 3-1 Soil hydraulic parameters measured by pressure plate and inversion

Estimated method	Soil layer (cm)	θ_r	θ_s	α	n	K_s
Measured by pressure plate	0–20	0.129	0.492	0.014	1.571	–
	20–60	0.13	0.433	0.007	1.846	–
	60–100	0.123	0.515	0.017	1.560	–
Inversion result	0–20	0.1143	0.417	0.0059	1.625	10.58
	20–60	0.0866	0.494	0.0077	1.765	24.85
	60–100	0.0546	0.500	0.0052	1.870	74.66

Notes: θ_r is the residual water content ($cm^3 \cdot cm^{-3}$); θ_s is the saturated water content ($cm^3 \cdot cm^{-3}$); K_s is the saturated hydraulic conductivity ($cm \cdot day^{-1}$); α (cm^{-1}) and n (–) are empirical coefficients that affect the shape of the hydraulic functions.

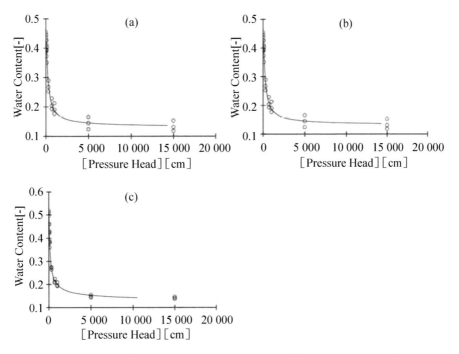

Figure 3-3 Measured soil water content at different pressure head
and the fitted soil water retention curves with RETC code (van Genuchten
et al., 1991) in (a) 0-20 cm; (b) 20-60 cm; (c) 60-100 cm

3.3.5.3 Root distribution

Considering vertical differences in maize root density were stronger than the horizontal differences, and differences in root density among inter-row positions were relatively small (Anderson, 1987; Anderson, 1988; Liedgens and Richner, 2001; Qin et al., 2006), only vertical differences in maize root density were taken into account, and maize root distribution was calculated with model proposed by Vrugt et al. (2001):

$$\beta(z) = \left[1 - \frac{z}{z_m}\right] e^{\frac{P_z}{z_m}|z^* - z|} \qquad (3-11)$$

where z_m was the maximum rooting depth in the vertical direction; z^* was an empirical parameter to describe the location of the maximum water uptake in the vertical direction; and P_z was an empirical parameter describing the non-symmetrical root geometry in the vertical direction. Considering the sowing depth, z^* was set as 10 cm. Since most of the maize root system was concentrated in the 0-60 cm depth (Asadi et al., 2002; Zhou et al., 2008; Gheysari et al., 2009), the depth of the root zone was set as 70 cm in this study. P_z was set to 1.0 according to Vrugt et al. (2001). Gao et al. (2014) found that although the plastic mulching system resulted in a greater root density than the un-mulched system, it did not alter the relative vertical distribution of maize roots. Therefore, the same parameters of the root density distribution were used for NM and PM systems in this study. The root distribution model will be applied in Eq. 8 and Eq. 5 to calculate the spatial distribution of root water uptake.

3.3.6 Statistical analysis

For each soil layer (0–20 cm, 20–60 cm and 60–100 cm), the average volumetric soil water content was calculated using the observed values. The middle points of each layer in the left boundary and right boundary (Figure 3-2) were selected as the observation points. Peformance of Hydrus-2D was evaluated via the root mean square error (RMSE) and the relative root mean square error (RRMSE) (Feng et al., 2017):

$$\text{RMSE} = \sqrt{\frac{\sum_{i=1}^{n}(SWC_{obs,i} - SWC_{model,i})^2}{n}} \qquad (3-12)$$

$$\text{RRMSE} = \text{RMSE}/SWC_{obs,mean} \qquad (3-13)$$

where $SWC_{obs,i}$ was the observed soil water content at time i, and $SWC_{model,i}$ was the modeled soil water content at time i, and n was the number of measurement times, $SWC_{obs,mean}$ was the mean value of the observed soil water content.

With Hydrus-2D, we assessed the 2D distribution of soil water daily, and we computed the average soil water content for the whole growing season as the average value of the daily results. The 2D distribution of the average soil water content in PM minus that in NM was calculated as the difference between PM and NM. All the calculations were completed using Matlab R2015b (MathWorks Inc., USA).

In this study, we adopted the relationship of stem-flow/through-fall and incident rainfall proposed by Martello et al. (2015) which was created in a sub-humid area with annual rainfall distributed fairly uniformly throughout the year. However, under different climate conditions, the parameter a and b in Eq. 10 and 11 might be different. Therefore we analyzed the sensitivity of soil water content to parameters a and b. The sensitivity coefficient of a or b was calculated during the growing season of 2015 using a single factor sensitivity analysis method that was reported in Liu et al. (2013). Normally, the practice of sensitivity analysis was to change parameters by 1% to obtain the sensitivity coefficient, however, as suggested in Liu et al. (2013), the magnitude of variation of a or b was set here at 10% to avoid possible disturbances associated with the numerical solving process.

3.4 Results

3.4.1 Comparison of the performance of different treatment methods for rainfall infiltration in PM

Figure 3-4 shows the simulated and measured SWC in different soil layers and positions, and

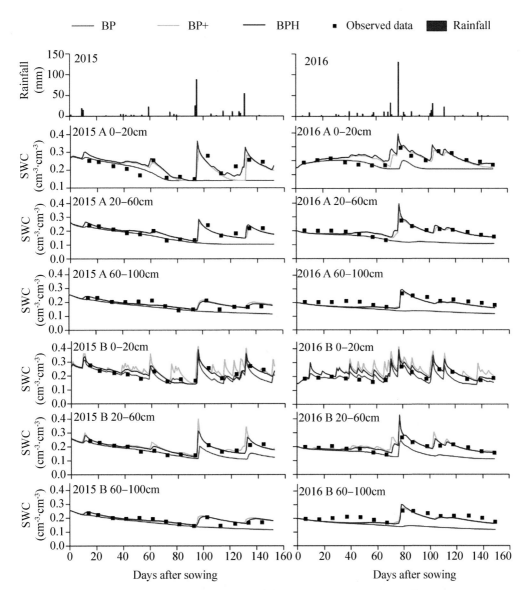

Figure 3-4 Performance of BP, BP+ and BPH in reproducing soil water dynamic in the different soil layers of field with plastic mulching.
(A) at the middle of plastic strip; (B) at the middle of bare strip

Table 3-2 shows the corresponding RMSE and RRMSE. The BP scenario did not reproduce the water content peaks caused by concentrated rainfall at the middle of plastic strip in both years. The performances improved with depth, but in general the RRMSEs were poor (>0.3, Table 3-2) and therefore this scenario was unsuitable to represent the field context. At the middle of bare strip, simulation BP was doing better than under the plastic strip in 2015 (the year with rainfall lower than the historic mean) with RMSEs ranging from 0.029 to 0.035

$cm^3 \cdot cm^{-3}$, and RRMSEs ranging from 0.150 to 0.190 in different soil layers. However, in 2016, a year with rainfall closer to the historic mean, the RMSE in 20-60 cm and 60-100 cm at the middle of bare strip were 0.043 $cm^3 \cdot cm^{-3}$ and 0.070 $cm^3 \cdot cm^{-3}$, and the RRMSE was 0.224 and 0.334 respectively, which were not convincing. The soil water content was underestimated in those two layers in 2016.

Table 3-2 The root mean square error (RMSE) and relative root mean square error (RRMSE) for the simulation with BP, BP+ and BPH in plastic mulch field in year 2015 and 2016

Treatments	Positions	Soil Depth (cm)	2015		2016	
			RMSE ($cm^3 \cdot cm^{-3}$)	RRMSE	RMSE ($cm^3 \cdot cm^{-3}$)	RRMSE
BP	Middle of plastic strip	0-20	0.065	0.302	0.062	0.301
		20-60	0.062	0.325	0.071	0.366
		60-100	0.034	0.183	0.071	0.338
	Middle of bare strip	0-20	0.032	0.150	0.025	0.124
		20-60	0.035	0.190	0.043	0.224
		60-100	0.029	0.161	0.070	0.334
BP+	Middle of plastic strip	0-20	0.035	0.165	0.027	0.131
		20-60	0.019	0.101	0.016	0.083
		60-100	0.018	0.099	0.024	0.112
	Middle of bare strip	0-20	0.030	0.142	0.056	0.281
		20-60	0.023	0.123	0.022	0.114
		60-100	0.018	0.097	0.023	0.110
BPH	Middle of plastic strip	0-20	0.032	0.151	0.031	0.149
		20-60	0.020	0.105	0.018	0.092
		60-100	0.017	0.089	0.025	0.119
	Middle of bare strip	0-20	0.022	0.104	0.032	0.163
		20-60	0.020	0.108	0.023	0.118
		60-100	0.016	0.086	0.026	0.123

Under the mulched strip, the difference between scenarios BP+ and BPH was minimal: BP+ resulted in a slightly lower soil water content than BPH in the upper soil layer in 2015 and 2016. Deeper down, the differences became small. The RRMSEs ranged from 0.083 to 0.165 in different soil layers over these two years for BP+, ranged from 0.089 to 0.151 for BPH. Both scenarios could therefore be considered adequate under the mulched strip. Nonetheless, the two scenarios yielded different results in the bare strip, where BP+ produced higher soil water content in 0-20 cm than BPH in both years (Figure 3-4). In 2016, this led to a clear overestimation of SWC in 0-20 cm (Table 3-2). We can conclude that the BPH simulation yielded the best results in 2015 and 2016 at the middle of bare strip with RMSEs ranging from 0.016 to 0.032 $cm^3 \cdot cm^{-3}$ and RRMSEs ranging from 0.086 to 0.163 in the different soil layers. Deviation between observed and simulated values may be caused by heterogeneous soil properties and irregular root distribution. Although different soil hydraulic parameters were applied in three soil layers, the heterogeneity of soil properties in each soil layer was not considered. At the same time, in the actual production, root distribution maybe quite irregular as a result of heterogeneity of soil properties. We were not able to take these possible effects into account due to the lack of data. Nevertheless, our results suggest BPH has better performance compared to BP and BP+, and the error seems acceptable in view of soil heterogeneity and irregular root distribution.

For the BPHscenario, the sensitivity of soil water content to the change of rainfall redistribution parameter a and b (Eq. 10 and 11) was shown in Table 3-3. 10% change of parameter a and b resulted in 0.003-0.108%, 0.002-0.239% change of soil water content in different soil layers and positions. The sensitivity of soil water content to a and b declined with increasing soil depth. Those results indicate the soil water content has very low sensitivity to the parameter a and b.

Table 3-3 Sensitivity of soil water content to the 10% change of parameters a and b in the rainfall redistribution model for BPH simulation

Parameters	Soil layers (cm)	Positions	
		Mulch strip	Bare strip
a	0-20	0.108%	0.099%
	20-60	0.095%	0.080%
	60-100	0.003%	0.039%
b	0-20	0.132%	0.239%
	20-60	0.095%	0.108%
	60-100	0.002%	0.034%

3.4.2 *Predicted field-scale spatial variation of soil water content in NM and PM*

As BPH performed better to reproduce our field data than BP and BP+, we used only this strategy to study the spatial soil water distribution in NM and PM. Figure 3-5 shows the simulated 2D distribution of the average soil water content under NM and PM treatments and their difference (ΔSWC). The simulation indicated that the top horizon (0-20 cm) contained more water than deeper in the profile for both years, and the water distribution in the top horizon showed high spatial variation. In PM, the bare strip was drier on average than the plastic strip, and the highest average SWC occurred near the planting hole (labelled with ③ and ④ in Figure 3-5). The centre zone of plastic strip (labelled with ① and ② in Figure 3-5) was drier than the zone near the planting hole and the bare strip was direst.

The ΔSWC ranged from 0.011 $cm^3 \cdot cm^{-3}$ to 0.043 $cm^3 \cdot cm^{-3}$ for the 2015 water conditions and from 0.007 $cm^3 \cdot cm^{-3}$ to 0.038 $cm^3 \cdot cm^{-3}$ for 2016. The positive influence of PM on the soil

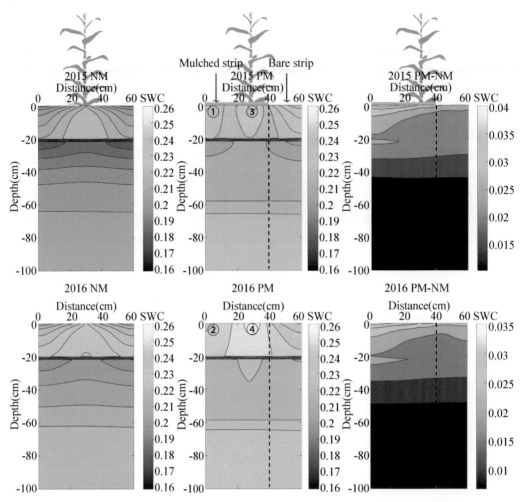

Figure 3-5 Two-dimensional average soil water content (SWC, cm³·cm⁻³) distribution during the growing seasons of 2015 and 2016 for the partial plastic mulching treatment (PM) and the no mulch treatment (NM), and the comparison of their differences (PM-NM)

moisture availability was higher in the top layer than in deeper layers. The ΔSWC in the centre of the plastic strip was highest, and it decreased towards the planting holes and bare strip. PM improved the SWC not only in the plastic strip but also in the bare strip.

3.5 Discussion

Our results suggest that the performance of Hydrus-2D in rain-fed field with plastic mulching depends strongly on the way the incoming rainfall is partitioned over the different types of surface boundaries and whether or not canopy redistribution is taken into account. Previous researches in arid regions suggested that Hydrus-2D could simulate the soil moisture dynamics well without taking into account these processes (Han et al., 2015; Li et al., 2015; Liu et al., 2013), probably because in these regions rainfall amounts are much smaller and drip irrigation is dominating the soil water dynamics under the plastic sheet. However, in rain-fed semi-arid areas, rainfall is the main soil water source and neglecting film side infiltration obviously leads to a reduction of the amount of water entering in the soil. In our case, about 67% (equal to 225 mm in 2015 and 253 mm in 2016) of incoming water would be omitted if the film side infiltration is not taken into account, explaining why the performance of the BP simulations is unsatisfactory.

The difference betweenBP+and BPH is more subtle, since the amount of rainfall is the same for both scenarios, but the infiltration location is different: in BP+all rainfall goes to the bare strip, whereas in BPH part of it infiltrates in the planting holes. Rain-fed semi-arid area is usually characterized by high wetting-drying frequency because of limited rainfall and the lack of irrigation (Borken and Matzner, 2009). While the BP+led to obvious stronger wetting intensity compared to BPH (Figure 3-4). This is the main reason why BP+simulation is not as good as BPH in rain-fed semi-arid area. Canopy redistribution is an important process in rainfall infiltration in maize field. Martello et al. (2015) observed that 78% of the rainfall was intercepted by the maize leaves under closed canopy conditions and transferred along the stem into soil and only 22% reached the ground directly. Paltineanu and Starr (2000) also obtained similar results. Our results suggest it is absolutely necessary to take rainfall canopy redistribution into consideration to simulate the soil water dynamic in rain-fed field with plastic mulching.

Our simulationwith BPH shows that the highest SWC occurs in the zone near the planting hole, the center zone of the plastic strip was drier, and the bare strip was the driest. Lowest soil moisture levels in the bare strip probably due to a higher exposure to evaporation. The lower soil moisture at the middle of plastic strip results from the fact that the plastic cover impedes rainfall infiltration. Highest soil moisture in zone near planting hole resulted from the stem-flow which improved the rainfall infiltration and the effects of plastic mulch which reduced the evaporation. The simulation also suggested that PM not only improved the average SWC in the plastic strip, but also remarkably improved the SWC in the bare strip as compared to NM. This is a result of lower evaporation in PM and soil water transport between the plastic and bare strips. In PM, the evaporation is largely reduced, which then leads to higher soil water content. During the drying stage, evaporation first leads to a reduction in soil moisture in the bare strip; then, because of the existing water potential gradient between the mulched strip and the bare strip, water is transported from the plastic strip to the bare strip, which partly compensates for the water loss in the bare strip. The improved soil moisture status in the bare strip of PM should be taken into account to evaluate its agronomical and environmental effects.

Nevertheless, in this study we only aimed at representing partial plastic mulching with PE material and following the common plastic-bare strip pattern as widely available on the Loess Plateau and in the Northeast of China. However, new plastic mulching methods, such as two ridges and furrow plastic mulching (Zhou et al., 2009), as well as new biodegradable materials (Saglam et al., 2017; Kasirajan et al., 2012) are currently being developed. Especially for plastic mulching with biodegradable materials, some extra processes should be taken into account, since these plastics decompose faster and present cracks changing the infiltration processes. In this case, a no-flux boundary which is constant in time may not be suitable for the mulched strip.

3.6 Conclusion

In the present study, we tested the performance of three strategies to represent rainfall infiltration on rain-fed fields with partial plastic mulching and applied the optimized strategy to investigate the spatial variation of soil moisture. The comparison demonstrated that when rainfall canopy redistribution and film side infiltration were neglected, Hydrus-2D failed to reproduce the soil water dynamics in all soil layers in the plastic strip and in the deep soil layers (20–60 cm and 60–100 cm) in the bare strip as a result of underestimation of rainfall infiltration. Integration of the process of film side infiltration by increasing the rainfall amount in bare strip results in too wet conditions in the bare strip as a result of overestimation of rainfall infiltration in bare strip. Hydrus-2D shows good performance when both rainfall canopy redistribution and film side infiltration are taken into account. This suggests that rainfall redistribution and film side infiltration are not negligible and should be correctly added as boundary conditions in the numerical solution of the Richards equation in rain-fed field with plastic mulching.

Our simulations with the optimized strategy suggest that the highest SWC appears in the zone near the planting hole, the center zone of the plastic strip is drier, and the bare strip is the driest. Compared to the NM, PM not only improves the SWC in the plastic strip but also improves the SWC in the bare strip. This implies to measure soil moisture dynamics and soil processes driven by soil moisture, difference between PM and NM should not only compared in plastic strip, but also should be compared in the zone near planting hole and in the bare strip due to different soil moisture conditions among those locations.

3.7 Reference

Allen R. G., Pereira L. S., Raes D., et al, 1998. FAO 56 Irrigation and drainage paper: crop evapotranspiration. Food and Agriculture Organization [M]. Rome: Food and Agriculture Organization.

Anderson E. L, 1987. Corn root-growth and distribution as influenced by tillage and nitrogen-fertilization [J]. Agron. J., 79: 544–549.

Anderson E. L, 1988. Tillage and n-fertilization effects on maize root-growth and root-shoot ratio [J]. Plant Soil, 108: 245-251.

Asadi M. E., Clemente R. S., Das Gupta A., et al, 2002. Impacts of fertigation via sprinkler irrigation on nitrate leaching and corn yield in an acid-sulphate soil in Thailand [J]. Agr. Water Manage., 52: 197-213.

Borken W., and Matzner E, 2009. Reappraisal of drying and wetting effects on C and N mineralization and fluxes in soils [J]. Global Change Biol., 15: 808-824.

Chen B., Yan C., Garré S., et al, 2018. Effects of a 'one film for 2 years' system on the grain yield, water use efficiency and cost-benefit balance in dryland spring maize (Zea mays L.) on the Loess Plateau, China [J]. Arch. Agron. Soil Sci., 64: 939-952.

Dane J. H., and Hopmans J. W, 2002. Pressure plate extractor. Methods of soil analysis. Part 4. Physical methods [M]. SSSA, Madison.

Deng X. P., Shan L., Zhang H., et al, 2006. Improving agricultural water useefficiency in arid and semiarid areas of China [J]. Agric. Water Manage., 80: 23-40.

Dong H., Li W., Tang W., et al, 2009. Early plastic mulching increases stand establishment and lint yield of cotton in saline fields [J]. Field Crop. Res., 111: 269-275.

Dusek J., Ray C., Alavi G., et al, 2010. Effect of plastic mulch on water flow and herbicide transport in soil cultivated with pineapple crop: a modeling study [J]. Agric. Water Manage., 97: 1637-1645.

Falkenmark M., and Rockström J, 2004. Balancing water for humans and nature [M]. Earthscan Publications, London.

FAO, 2006. Guidelines for soil description [M]. Rome: Food and Agriculture Organization.

Feddes R. A., Kowalik P. J., and Zaradny H, 1978. Simulation of field water use and crop yield [M]. John Wiley and Sons, New York.

Feng Y., Cui N., Gong D., et al, 2017. Evaluation of random forests and generalized regression neural networks for daily reference evapotranspiration modelling [J]. Agr. Water Manage., 193: 163-173.

Filipovic V., Romic D., Romic M., et al, 2016. Plastic mulch and nitrogen fertigation in growing vegetables modify soil temperature, water and nitrate dynamics: Experimental results and a modeling study [J]. Agr. Water Manage., 176: 100-110.

Fisher P. D, 1995. An alternative plastic mulching system for improved water management in dryland maize production [J]. Agr. Water Manage., 27: 155-166.

Gao Y. H., Xie Y. P., Jiang H. Y., et al, 2014. Soil water status and root distribution across the rooting zone in maize with plastic film mulching [J]. Field Crop. Res., 156: 40-47.

Gheysari M., Mirlatifi S. M., Homaee M., et al, 2009. Nitrate leaching in a silage maize field under different irrigation and nitrogen fertilizer rates [J]. Agr. Water Manage., 96: 946-954.

Gong D., Hao W., Mei X., et al, 2015. Warmer and wetter soil stimulates assimilation more than respiration in rainfed agricultural ecosystem on the China Loess Plateau: the role of partial plastic film mulching tillage [J]. Plos One, 10, e0136578.

Gong D. Z., Mei X. R., Hao W. P., et al, 2017. Comparison of ET partitioning and crop coefficients between partial plastic mulched and non-mulched maize fields [J]. Agr. Water Manage., 181: 23-34.

Han M., Zhao C. Y., Feng G., et al, 2015. Evaluating the effects of mulch and irrigation amount on soil water distribution and root zone water balance using HYDRUS-2D [J]. Water, 7: 2622-2640.

Kader M. A., Senge M., Mojid M. A., et al, 2017. Recent advances in mulching materials and methods for modifying soil environment [J]. Soil Till. Res., 168: 155-166.

Kasirajan S., and Ngouajio M, 2012. Polyethylene and biodegradable mulches for agricultural applications: a review [J]. Agron. Sustain. Dev., 32: 501-529.

Li X. Y., Shi H. B., Simunek J., et al, 2015. Modeling soil water dynamics in a drip-irrigated intercropping field under plastic mulch [J]. Irrig. Sci., 33: 289-302.

Liedgens M., and Richner W, 2001. Minirhizotron observations of the spatial distribution of the maize root system [J]. Agron. J., 93: 1097-1104.

Liu M. X., Yang J. S., Li X. M., et al, 2013. Numerical simulation of soil water dynamics in a drip irrigated cotton field under plastic mulch [J]. Pedosphere, 23: 620-635.

Liu Q., Chen Y., Li W., et al, 2016. Plastic-film mulching and urea types affect soil CO_2 emissions and grain yield in spring maize on the Loess Plateau, China [J]. Sci. Rep., 6: 28150.

Martello M., Dal Ferro N., Bortolini L., et al, 2015. Effect of incident rainfall redistribution by maize canopy on soil moisture at the crop row scale [J]. Water, 7: 2254-2271.

NBSC (National Bureau of Statistics of China), 2016. China statistical yearbook 2016. China Statistics Press, Beijing.

Paltineanu I. C., and Starr J. L, 2000. Preferential water flow through corn canopy and soil water dynamics across rows [J]. Soil Sci. Soc. Am. J., 64: 44-54.

Qin R. J., Stamp P., and Richner W, 2006. Impact of tillage on maize rooting in a Cambisol and Luvisol in Switzerland [J]. Soil Till. Res., 85: 50-61.

Qin W., Hu C. S., and Oenema O, 2015. Soil mulching significantly enhances yields and water and nitrogen use efficiencies of maize and wheat: a meta-analysis [J]. Sci. Rep., 5: 16210.

Ren X., Zhang P., Liu X., et al, 2017. Impacts of different mulching patterns in rainfall-harvesting planting on soil water and spring corn growth development in semihumid regions of China [J]. Soil Res., 55: 285-295.

Richards L, 1931. Capillary conduction of liquid in porous media [J]. Physics, 1: 318-333.

Ritchie J. T, 1972. Model for predicting evaporation from a row crop with incomplete cover [J]. Water Resour. Res., 8: 1204-1213.

Rockström J., Karlberg L., Wani S. P., et al, 2010. Managing water in rainfed agriculture-the need for a paradigm shift [J]. Agric. Water Manage., 97: 543-550.

Rost S., Gerten D., Bondeau A., et al, 2008. Agricultural green and blue water consumption and its influence on the global water system [J]. Water Resour. Res., 44: 137-148.

Saglam M., Sintim H. Y., Bary A. I., et al, 2017. Modeling the effect of biodegradable paper and plastic mulch on soil moisture dynamics [J]. Agr. Water Manage., 193: 240-250.

Šimůnek J., Vogel T., and van Genuchten M. T, 1994. The swms_ 2d code for simulating water flow and solute transport in two-dimensional variably saturated media [M]. Riverside: U. S. Salinity Laboratory Agricultural Research Service, USDA.

Tarara J. M, 2000. Microclimate modification with plastic mulch [J]. HortScience, 35: 169-180.

van Genuchten M. T., Leij F. J., and Yates S. R, 1991. The RETC code for quantifying the hydraulic functions of unsaturated soils [M]. Robert S. Kerr Environmental Research Laboratory.

van Genuchten M. T, 1980. A closed-form equation for predicting the hydraulic conductivity of unsaturated soils [J]. Soil Sci. Soc. Am. J., 44: 892-898.

Vrugt J. A., Hopmans J. W., and Simunek J, 2001. Calibration of a two-dimensional root water uptake model [J]. Soil Sci. Soc. Am. J., 65: 1027-1037.

Vereecken H., Schnepf A., Hopmans J. W., et al, 2016. Modeling soil processes: review, key challenges, and new perspectives [J]. Vadose Zone J., 15: 1-57.

Wu Y., Huang F., Jia Z., et al, 2017. Response of soil water, temperature, and maize (Zea may L.) production to different plastic film mulching patterns in semi-arid areas of northwest China [J]. Soil Till. Res., 166: 113-121.

Xiao Q., Zhu L. X., Shen Y. F., et al, 2016. Sensitivity of soil water retention and availability to biochar addition in rainfed semi-arid farmland during a three-year field experiment [J]. Field Crop. Res., 196: 284-293.

Zhao H., Xiong Y. C., Li F. M., et al, 2012. Plastic film mulch for half growing-season maximized WUE and yield of potato via moisture-temperature improvement in a semi-arid agroecosystem [J]. Agr. Water Manage., 104: 68-78.

Zhou L. M., Li F. M., Jin S. L., et al, 2009. How two ridges and the furrow mulched with plastic film affect soil water, soil temperature and yield of maize on the semiarid Loess Plateau of China [J]. Field Crop. Res., 113: 41-47.

Zhou S. L., Wu Y. C., Wang Z. M., et al, 2008. The nitrate leached below maize root zone is available for deep-rooted wheat in winter wheat-summer maize rotation in the North China Plain [J]. Environ. Pollut., 152: 723-730.

4

Field-scale Spatial Variation of Soil Moisture Fluctuation

4.1 Abstract

Extreme events in drylands lead to a high risk of soil degradation. To adjust intense SMF in agricultural systems, it is necessary to evaluate the influence of different agricultural practices on SMF. Partial PM is a widely used practice in drylands. Previous research typically studied the effects of PM on SMF using point-scale measurements. However, the results were inconclusive, as opposite effects may exist in different positions of the soil profile and soil surface. In order to take this spatial heterogeneity and the complexity of soil water fluxes under PM into account, we performed ERT measurements and simulated field conditions using Hydrus 2D. The results indicate that the application of PM results in contrasting SMF between the mulched strip and the bare strip in PM. The soil moisture in the mulched strip was relatively stable compared to that in the bare strip. The difference in SMF between the mulched strip and the bare strip was predominant in the surface layers. Nevertheless, while comparing the SMF between PM and NM using Hydrus 2D, we found that PM reduced the SMF for the whole soil profile compared to NM. This means that PM not only alleviated SMF in the mulched strip but also in the bare strip. Therefore, we concluded that PM is an effective practice to reduce the impact of intense SMF in rain-fed, semi-arid areas, leading to less soil degradation.

Keywords: Soil moisture fluctuation; Hydrus-2D; Electrical resistivity tomography; Spatial variation.

4.2 Introduction

Arid, semi-arid, and seasonally arid areas occupy about one-third of the Earth's land. These regions usually experience intense SMF due to the long duration of droughts and subsequent intense wetting pulses, especially when irrigation is not available (i.e., in rain-fed areas) (Schimel et al., 2007; Manzoni et al., 2014). Considering the increasing probability of extreme events due to climate change, extreme events are likely to occur even more frequently in the future (Donat et al., 2016; Borken et al., 2009; Solomon, 2007). It is well documented that intense SMF can reduce the stability of aggregates and accelerate the turnover of soil organic matter, both of which lead to a high risk of soil degradation and subsequent limitations in terms of soil productivity in drylands (Birch, 1958; Jarvis et al., 2007; Turner et al., 2001; Borken et al., 2009; Chepkwony et al., 2001). In agricultural systems, SMF is not only driven by climate conditions but also by agricultural management practices. To reduce strong SMF in agricultural systems, it is necessary to thoroughly understand the influence of these practices on soil moisture dynamics.

Plastic mulching (PM) is an agricultural practice that alleviates the effects of drought on crop production worldwide (Kasirajan and Ngouajio, 2012; Scarascia-Mugnozza and Russo,

2011). In China, PM has been adopted in nearly 20% of farmland (NBSC, 2014). The influence of PM on soil water dynamics has been widely researched in various regions and crop cultivation systems (Gan et al., 2013; Seyfi et al., 2007; Lordan et al., 2015). Previous research reported that PM is effective in reducing soil water evaporation (Li et al., 2013a; Zhou et al., 2009; Li et al., 2013b), forming a relatively stable soil moisture content and buffering large fluctuations in soil moisture (Kader et al., 2017; Feng et al., 2014; Ghosh et al., 2006). However, most of these studies were carried out using soil water measurement methods, such as time-domain reflectometry and soil sampling, while neglecting spatial variations in soil moisture dynamics.

To answer the question of whether PM is beneficial to alleviate SMF in rain-fed drylands, the evidence from point-scale measurements does not suffice. Fields with PM are usually composed of a combination of bare and mulched strips and planting holes within the mulched strip (Berger et al., 2013; Li et al., 2017; Zhao et al., 2014). If PM only reduces the SMF in specific locations (such as in the mulched strip) but strengthens SMF in others (such as the bare strip), the overall effect of PM on SMF may not be what we would expect in looking at point-scale measurements. The spatial variability of water fluxes has to be taken into account in this case because of the complex interaction among weather conditions, crop development, soil characteristics, and plastic film.

Hydrological modeling and geophysical measurements are two ways to obtain information on field-scale spatial variations in soil water dynamics. Modeling tools, such as Hydrus 2D, are a fast and low-cost way to extrapolate the field-scale spatio-temporal patterns of soil moisture in PM systems (Šimůnek et al., 2008). However, the obtained results are based on a series of assumptions in terms of root distribution, soil hydraulic parameters, and transport boundaries (Šimůnek et al., 2008), which limits the reliability of these models in complex cases. Geophysical measurement techniques, and, more specifically, ERT, allow for the monitoring of 2D soil moisture dynamics quasi non-invasively and with a relatively high spatial resolution (Whalley et al., 2017; Vanderborght et al., 2013). As equipment has improved over the last decades, the monitoring potential of these methods has become powerful; thus, the acquisition of time-lapse data with a high temporal resolution is possible (Garré et al., 2011; Beff et al., 2013). Considering the complexity of soil water transport processes in PM fields, combining geophysical and modeling tools is beneficial to improve the reliability of results.

The objectives of this study were ①to investigate how weather conditions and mulching practices together generate spatial variations in SMF in PM fields and ②to quantify and explain the differences in SMF between fields with and without PM in a 2D soil profile. We assumed that PM is an effective practice to alleviate SMF in drylands and causes vertical and horizontal patterns in the soil moisture dynamics. A maize field located on the Loess Plateau in China was selected to validate this assumption during the crop growing season of 2017. We used field and modeling data to investigate the effect of PM on spatial variations in SMF.

4.3 Materials and Methods

4.3.1 *Research site*

The research site was located in Shouyang, Shanxi Province, China (37°45′N, 113°12′E, 1080-m altitude). The climate is semi-arid according to the UNEP's classification system (UNEP, 1992). Under average climatic conditions, the area receives 480 mm of precipitation annually, about 70% of which occurs in the summer from June until September. The conventional cropping system is continuous maize cultivation. Usually, maize is sown in late April or early May and is harvested in late September or early October. The soil texture is defined as a loam under the USDA's classification system and is classified as calcaric Cambisol according to the World Reference Base for Soil Resources (FAO, 2006).

4.3.2 *Layout for field with and without plastic mulching*

As shown in Figure 4-1, the field with PM included mulched strips with a width of 80 cm, bare strips with width of 40 cm, and maize planting holes. The soil was covered by a clear and impermeable PE film with a 10-μm thickness. Rotary tillage and fertilizer applications were carried out before PM. In accordance with local practice, fertilizers were applied at rates of N 225 kg·ha^{-1} (urea), P_2O_5 162 kg·ha^{-1} (calcium superphosphate), and K_2O 45 kg·ha^{-1} (potassium chloride) before sowing without topdressing. After then, the spring maize was sown manually with the row distance of 60 cm and an in-row plant spacing of 30 cm (sowing density = 55,556 plants·ha^{-1}). For the field without PM, the same tillage, fertilization, and sowing practices were applied, but no plastic film was adopted. For more field management information, please see Chen et al. (2018).

4.3.3 *2D simulation with Hydrus-2D*

The simulation of soil moisture dynamics under PM was carried out using Hydrus 2D (Šimůnek et al., 2008). Hydrus 2D uses the Galerkin finite element method to numerically solve the governing equation (Richards, 1931) in order to determine the variably saturated water flow and Mualem-van Genuchten relationships to estimate the soil water retention curve and the unsaturated hydraulic conductivity function (van Genuchten, 1980). We adopted the Feddes model (Feddes et al., 1978) to describe the root water extraction as a function of soil water potential in the root zone.

During the growing season of 2017, meteorological data, including air temperature, air humidity, wind speed, sunshine duration, and rainfall, were recorded hourly by an on-site, automatic weather station (Campbell Scientific Inc., Logan, UT, USA). At the same time, the leaf area was monitored following the method of Li et al. (2008). With the collected mete-

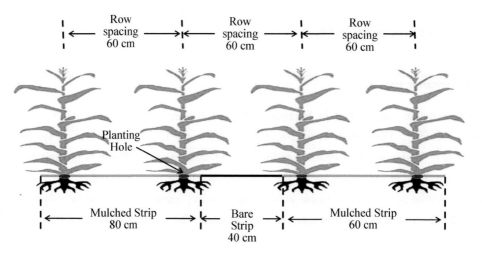

Figure 4-1 Layout of maize field with plastic mulching

orological data, the reference crop evapotranspiration (ET_0) was computed using the Penman-Monteith equation, and then the crop evapotranspiration was calculated by multiplying ET_0 by the crop coefficient (K_c) (Allen et al., 1998). In this study, the local measured Kc, which was reported by Chen et al. (2018) and Gong et al. (2017) was used. Potential evaporation and transpiration were calculated separately from crop evapotranspiration using Beer's law, which partitions the solar radiation component of the energy budget via interception by the leaf area index (Ritchie, 1972). The soil hydraulic parameters were determined in 2015 and 2016, as reported in Chapter 3. We assumed a uniform maize root distribution in the horizontal direction and described the non-symmetrical root geometry in the vertical direction using the model of Vrugt et al. (2001). The root non-symmetrical parameter was set to 1.0, and the location of the maximum water was fixed at 10 cm from the collar horizontally. The depth of the root zone was 70 cm in this study.

In PM, we used an upper boundary composed of different zones with different properties using BPH. A no-flux boundary was imposed on the plastic strip, and an atmospheric boundary was imposed on the bare strip. A 2-cm wide variable flux boundary was imposed in the planting hole to take the rainfall redistribution caused by the canopy into account. In NM, a similar 2-cm wide variable flux boundary was imposed in the sowing position, and other parts of the upper boundary were expressed as an atmospheric boundary. For both PM and NM, a no-flux boundary was used for the left and right vertical boundaries of the domain due to symmetry reason, and a free drainage boundary was used to represent the bottom boundary. For details on the calculation of rainfall infiltration in different boundaries, please see Chen et al. (2018).

4.3.4 *Electrical resistivity tomography*

During the growing season of 2017, ERT measurements were carried out from the 7th day after sowing (DAS) to the 138th DAS, with measurement intervals of 1-5 days. During the whole growing season, a total of 75 ERT measurements were performed. The ERT measurements were carried out using an earth resistivity meter (Lippman, Germany) and a 24 electrode switch box (Lippman, Germany), as shown in Figure 4-2. The stainless steel electrodes (length: 12 cm, diameter: 5 mm) were placed in the soil with a spacing of 20 cm before we applied the plastic film.

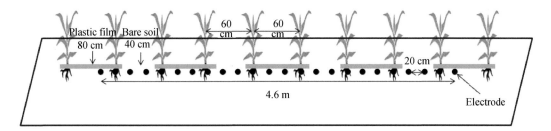

Figure 4-2 Scheme of the plastic mulching and layout of electrodes

We adopted a dipole-dipole array (Samouëlian et al., 2005) following the approach of Michot et al. (2003), in which the spacing between the electrodes of each dipole was subsequently increased to overcome the disadvantage of the dipole-dipole array in which increasing the overall length of the array and the depth of the investigation leads to a drop in the potential. In total, 7 combinations of measurement level (n) and electrodes spacing (a) were used, resulting in 638 quadrupoles (319 normal and 319 reciprocal measurements) (Figure 4-3).

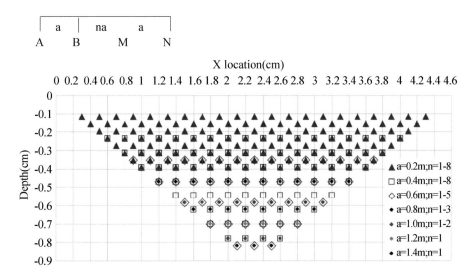

Figure 4-3 Measurement sequence using a dipole-dipole arrangement (picture established using GeoTest, Dr. A. Rauen, Germany)

We used the injected current, the measured potential, and the reciprocal error to filter the data set, and the reciprocal error ($\varepsilon N/R$) was calculated as follows:

$$\varepsilon_{N/R} = \mid EC_N - EC_R \mid /EC_{mean} \qquad (4-1)$$

where EC_N was the electrical conductivity obtained by normal measurement, EC_R was the resistivity obtained by reciprocal measurement, and EC_{mean} was the mean of EC_N and EC_R.

After data filtering, RES2DINV was used to invert the measured resistivities (Loke and Barker, 1996). We used a finite-difference forward modeling approach and a standard Gauss-Newton inversion scheme on regular mesh with four nodes per unit. The smoothness constraint was applied on the perturbation vector, and the Jacobian matrix was recalculated for all iterations. We used a damping factor (initial = 0.15, minimum = 0.05) to reduce the amplitude of the perturbation vector. For inversion model discretization, we used model cells with widths of half the unit spacing, as inversion model discretization usually gives optimum results for dipole-dipole arrays. A moderate layer thickness increased by 10% was adopted, and the number of model blocks was allowed to exceed the number of data points. We used an root mean square error (RMSE) of 2% as the convergence limit. A sequential time-lapse inversion was adopted, and the first data set was first inverted independently and was then used as a reference model for the inversion of later time steps.

The soil electrical conductivity (EC) obtained from the inversion was corrected to a standard temperature of 25℃ using the equation established by Keller and Frischknecht (1966):

$$EC_{25} = EC_T / [1 + \alpha (T-25)] \qquad (4-2)$$

where EC_{25} is the electrical conductivity at $T = 25$ ℃ and α equals $0.02℃^{-1}$. This means that a deviation of 1℃ in the temperature leads to a deviation of 2% in the electrical conductivity; EC_T is the electrical conductivity at temperature T.

We used the Waxman and Smits model (Waxman and Smits, 1968) to calibrate the EC-WC relationship:

$$EC_{bulk} = \frac{1}{\varphi} WC^n + EC_s \qquad (4-3)$$

where EC_{bulk} (S·m^{-1}) is the bulk electrical conductivity, WC is the water content, φ is ohmm, and n and ECs (S·m^{-1}) are the fitting parameters.

Three 5TE sensors (Decagon, USA) were installed at 10-cm, 40-cm, and 80-cm depths in the mulched strip to monitor the soil moisture, temperature, and electrical conductivity on an hourly basis in the three main soil horizons (0–20 cm, 20–60 cm, and 60–100 cm). This allowed us to establish a field-calibrated EC-WC relationship, covering the soil moisture range encountered during the growing season of the experiment.

4.3.5 Statistics

We compared the converted SWC obtained from ERT (i.e., SWC-ERT) with the SWC obtained by the 5TE sensors (SWC-sensors). The geophysical inversion was performed on a rectangular grid composed of 113 rectangular mesh cells. The grid was divided into seven layers with centers at 0.05-m, 0.15-m, 0.27-m, 0.4-m, 0.54-m, 0.69-m, and 0.86-m depths. The coordinates of the rectangular mesh center can be expressed as (X, Z), in which X represents the horizontal distance along the electrode line and Z represents the depth. The average values of SWC-ERT at (1.9, 0.05), (2.1, 0.05), (1.9, 0.15), and (2.1, 0.15) were compared with that of the SWC-sensors at a 0.1-m depth of the maize row. Then, the average values of SWC-ERT at (1.9, -0.4) and (2.1, -0.4) were compared with that of the SWC-sensors at a 0.4-m depth of the maize row, and the average values of SWC-ERT at (1.9, 0.69), (2.1, 0.69), (1.9, 0.86), and (2.1, 0.86) were compared with that of the SWC-sensors at a 0.8-m depth of the maize row. The RMSE was used to evaluate the performance of ERT in reproducing the soil moisture content.

We used coefficients of variation (CV) to quantify soil water fluctuation during the whole growing season (Beff et al., 2013; Zhao et al., 2017; Hu et al., 2011):

$$CV = \frac{\sqrt{[(SWC_{t1} - SWC_{mean})^2 + \cdots + (SWC_{tn} - SWC_{mean})^2]/n}}{SWC_{mean}} \quad (4-4)$$

where SWC_{ti} is the SWC at the $_{ith}$ ERT measurements or the $_{ith}$ simulated day (the simulated SWC was output daily), SWC_{mean} is the average of the SWC, and n is the total time of the ERT measurement or the simulated days.

A low CV means that the soil moisture was stable, while a high CV means that the soil water fluctuation was intense. The CV were calculated for each grid, which was produced by Hydrus 2D or ERT, and then the linear interpolation was employed to intercept the CV between adjacent grids. The statistics were conducted using MATLAB 2015b (MathWorks, USA).

4.4 Results and discussion

4.4.1 *Field-scale spatial variation of SMF in PM-Results from Hydrus-2D*

Using the calibrated and validated model obtained in Chen et al. (2018), we applied meteorological data from 2017 to simulate the field-scale spatio-temporal dynamics of soil moisture. Figure 4-4 shows the simulated spatial variations in SMF with Hydrus 2D in 2017 using the CV patterns of the SMF. The upper soil horizon is characterized by a high CV in the bare strip and near the planting hole, whereas the mulched strip is characterized by a relatively low CV. The variation decreased gradually with the depth up until the interfaces of the two soil horizons showed contrasting hydraulic properties. Near this interface, at a 20-30-cm depth, the CV was high and decreased toward the bottom of the second horizon (60 cm). At a 60-100 cm depth, the CV was relatively small. Thus, the spatial variations in SMF in PM can be generally conceptualized as follows: ①an active part with a high CV in the bare strip and zones near the planting hole of the surface layer and an inert part with a low CV in the mulched strip of the surface layer, ②a second layer below the surface layer characterized by a low CV, ③a third layer below the second layer with a high CV, and ④a bottom layer with a low CV. Logically, SMF not only depends on the location of the plastic mulched strip but also shows a non-linear relationship with soil depth due to the interactions between different soil horizons.

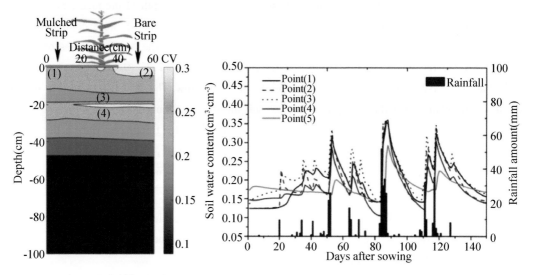

Figure 4-4 Simulated spatial variation of coefficients of variation (CV) in field with plastic mulching, and the dynamic of soil water content in selected points. Co-ordinate for point 1-5 was (2, -4), (58, -4), (30, -16), (30, -26) and (30, -50)

The different CV in different soil layers and parts can be explained by their sensitivity to rainfall, evaporation, and root water uptake. Figure 4-4 shows the simulated soil water dynamics in five selected points in zones with contrasting SMF in the simulated 2D soil profile. Because of the impermeability of plastic film, evaporation and rainwater availability are reduced in mulched soil (Li et al., 2013a; Zhou et al., 2009; Li et al., 2013b); in planting holes and bare strips, there is a more immediate response to rainfall and evaporation. Hence, it is logical that the mulched strip (represented by point 1) and the bare strip (represented by point 2) behave differently in terms of SMF. In addition, soil moisture was stable in the deepest soil layer (represented by point 5), as the influence of weather conditions was reduced and the root water uptake was small (as also shown by Zhao et al., 2017; Zhang et al., 2016; Beff et al., 2013).

As shown in Figure 4-4, the different behaviors in terms of SMF between points 3 and 4 are primarily caused by different responses to small rainfall. Compared to point 4, point 3 had a higher response to small rainfall. Small rainfall only wets a limited soil depth, and water from rainfall reaches point 4 less easily than point 3. Small rainfall and reduced wetting pulses after heavy rainfall could play an important role in keeping the soil moisture in point 3. At the same time, the response to drying in point 3 was less strong than in point 2 because of less contact with the atmosphere. Thus, point 3 had a higher and more stable soil moisture. Because heavier rainfall reached point 4, the soil moisture remained at a relatively low level before heavy rainfall; after heavy rainfall, there was a strong shift in the soil moisture.

4.4.2 Field-scale spatial variation of SMF in PM-Results from ERT measurement

4.4.2.1 Pedo-physical relationship

Figure 4-5-A shows that the sensors at 0-20 cm and 20-40 cm reacted to major rainfall events. However, some outliers were produced during these rainfall events (51-52 DAS, 70-71 DAS, and 83-87 DAS), which were discarded to fit the pedo-physical relationship (Figure 4-5-C). The produced outliers may have resulted from the arrival of fresh water during rainfall, the transport of salts, and re-equilibration between the pore water concentration and the soil grains (Rhoades et al., 1976; Revil et al., 2013). The arrival of fresh water during rainfall may reduce the salt concentration and lead to a low EC_{bulk}, while the

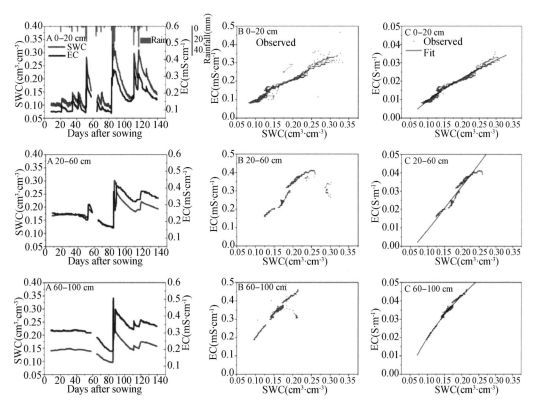

Figure 4-5 Soil water content (SWC) dynamics, soil bulk electrical conductivity at 25℃ (EC) obtained with the 5TE sensors and simplified Waxman and Smits model fits.
(A) Evolution of SWC and EC with time; (B) unfiltered SWC and EC data;
(C) filtered data and simplified Waxman and Smits model fit

transport of salts may reduce the EC_{bulk} in the upper layer and increase the EC_{bulk} in the lower layer. Re-equilibration between the pore water concentration and the soil grains during the wetting process improved the EC_{bulk}. As the simplified Waxman and Smits model did not take the changing pore salt concentration into account, these outliers were not used to fit the simplified Waxman and Smits model. The discarded data accounted for about 6% of the total measured data.

Generally, the fitting performance (Table 4-1) was satisfactory at all depths, even if the sigmoid form of the relationship in horizon 2 could not be represented by the simplified Waxman and Smits model. The sensitivity of EC to SWC increased with the soil depth (i.e., steeper slopes). Calibration relationships were used to convert the bulk EC distribution from ERT to SWC for each time step and were compared to the SWC obtained using the sensors. The residual errors of the model fitting in different soil layers are shown in Figure 4-6. There was a significant positive correlation between the measured SWC and residual errors in different soil layers.

Table 4-1 Fitted Parameters of the simplified Waxman and Smits (1968) model for each of the three horizons and the fitting performance

Horizon	φ (ohmm)	N (-)	ECs (S·m^{-1})	Number of observations	Coefficient of determination
0-20cm	10.58	0.60	-0.014	1424	0.97
20-60cm	3.91	1.08	-0.013	1424	0.92
60-100cm	5.53	0.40	-0.052	1424	0.98

4.4.2.2 The performance of ERT on soil water recovery

Generally, the agreement between the soil moisture obtained from ERT (SWC-ERT) and that obtained from the soil moisture sensors (SWC-sensors) were good (Figure 4-7). The RMSEs at 0-20 cm, 20-60 cm, and 60-100 cm were 0.020, 0.027, and 0.024 cm^3, respectively. Moreover, the SWC-ERT and SWC-sensors reacted similarly in terms of SMF except for the peak around 50 DAS. The CV of the SWC-ERT and SWC-sensors were 0.386 and 0.356, 0.188 and 0.212, and 0.174 and 0.224 at 0-20 cm, 20-60 cm, and 60-100 cm, respectively. Many factors influence electrical conductivity (e.g., temperature, pore water resistivity, surface electric conductivity, porosity), and the inversion problem is ill posed, causing uncertainty in the interpretation of the soil moisture values obtained using ERT (Brunet et al., 2010). Nevertheless, our results indicate that ERT measurements can retrieve the soil water dynamics in a satisfactory way using field-specific calibration, which is in agreement with the results of previous research (Brunet et al., 2010; Besson et al., 2010; Garré et al., 2012).

4.4.2.3 Spatial variation of SMF in PM

Figure 4-8 shows the patterns of SMF obtained using ERT in 2017. The conditions in the surface soil under the plastic sheet are rather constant, and, as expected, more variation was ob-

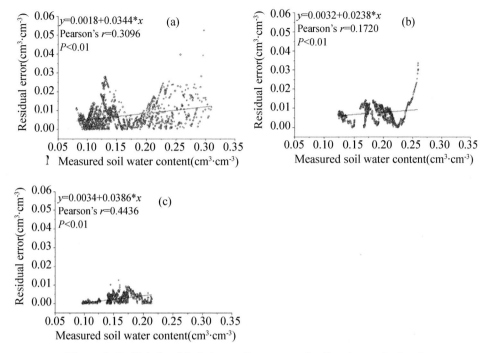

Figure 4-6　Relationship between the measured soil water content and the residual errors of the fitting of simplified Waxman and Smits (1968) model at (a) 0-20 cm, (b) 20-60 cm, and (c) 60-100 cm

served in the bare surface soil near the planting hole. At -0.2 m, there are two zones with higher variability. We also found an inert zone in the subsurface and in deep soil layers. Compared to the spatial patterns of SMF obtained using Hydrus 2D, the ERT measurements resulted in a more irregular distribution of zones with contrasting CV. The heterogeneity in the pattern of the SMF reflects the combined result of the natural heterogeneity of the soil hydraulic properties, root distribution, and rainfall infiltration. However, the ERT measurements confirm the main behavior recognized in the simulated profile (Figure 4-4), namely, that SMF is more stable in the surface soil layer in the mulched strip than in the bare strip.

4.4.3 *Comparison of the field-scale spatial variation of SMF between PM and NM*

In the vertical direction, NM and PM showed a similar behavior, in which the variation in SMF in the vertical direction can be summarized as a surface layer with a high CV, a second layer below the surface layer with a low CV, a third layer below the second layer with a high CV, and a deep soil layer with a low CV. In 2017, the CV ranged from 0.089-0.342 in PM with an average of 0.180 and from 0.105-0.381 in CV with an average of 0.201. The difference in the CV between the PM and NM ranged from -0.006 to -0.078 with an average of -0.021. For the whole soil section, the average CV was reduced by 10% under PM as compared

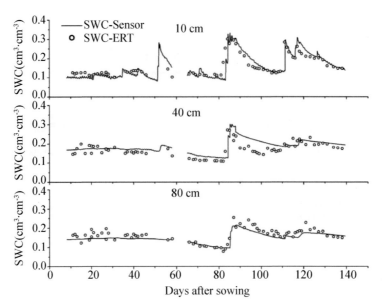

Figure 4-7 Performance of ERT in terms of soil water contentrecovery in field with plastic mulching. SWC-ERT is the soil moisture obtained by ERT, and SWC-sensors is the soil moisture obtained by the soil moisture sensors. At 0-20cm, 20-60cm, and 60-100 cm, SWC-ERT and SWC-sensors were compared in terms of the positions of the maize row and at depths of 10 cm, 40 cm, and 80 cm

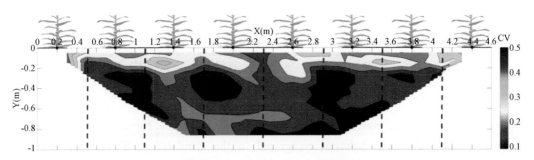

Figure 4-8 Patterns of soil moisture fluctuations, represented by the coefficient of variation of soil moisture (CV), throughout the season obtained using ERT in 2017. The zones between the two adjacent black dashed lines indicate the area simulated using Hydrus 2D in Figure 4-4 to allow for an easy comparison. The blue line on the x-axis indicates the location of the plastic film

to NM. The negative difference of CV between PM and NM indicated that PM not only reduced the CV in the mulched strip but also reduced the CV in the bare strip. Although the influence of PM on CV in the bare strip was not as large as that in the mulched strip, the soil drought degree was alleviated during the drying period in the mulched strip (points 3 and 5 in Figure 4-9).

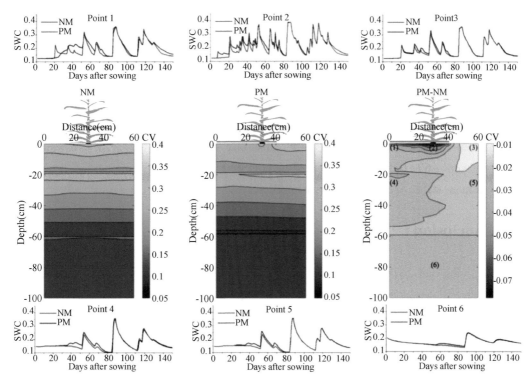

Figure 4-9 Spatial variation of soil moisture fluctuation in field with plastic mulching (PM), no-mulch filed (NM) in 2017 and their difference (PM-NM). CV: Coefficient of variation. PM-NM was calculated as the CV in PM minus the CV in NM. The coordinates of comparison points 1-6 are (2, -4), (30, -4), (58, -4), (2, -26), (58, -26), and (30, -80)

Generally, PM and NM differ most in the mulched strip (PM-NM, Figure 4-9). At the center of the mulched strip (represented by point 1 in Figure 4-9), the application of plastic film reduced the SMF peak value caused by rainfall in early growing stages. In the middle and at the end of the growing season, the SMF peaks were close to each other in NM and PM. In early stages, most rainfall infiltrates the soil without canopy redistribution (throughfall infiltration), while the application of plastic film largely reduces throughfall infiltration in the mulched strip. However, as the canopy develops, most rainfall is intercepted and then flows into the soil along the stem (stemflow infiltration) in both NM and PM. Thus, the influence of PM on throughfall infiltration was less predominant. This can explain why the influence of plastic film on SMF peaks was obvious in early stages but not in the middle or late growing seasons. At the same time, because of the lower evaporation in the mulched strip, the soil moisture remains higher in the middle or late growing seasons and fluctuates only in a narrow range. This leads to a lower CV for the whole growing season in the mulched strip. The lower SMF of PM in the zone near the planting hole (represented by point 2 in Figure 4-9) and at a 20-40 cm depth under the mulched strip (represented by point 4 in Figure 4-9) resulted from the low evaporative demand during the drying phases. In these zones, the difference between PM and NM is more predominant in early stages than in later stages. This is in agreement with previous research, suggesting that the effect of PM on evaporation reduces as the crop canopy develops. The lower SMF of PM in the bare strip (represented by points 3 and 5 in Figure 4-9) can be explained by water transport between the mulched and bare strips. During soil drying, because of the existing water potential gradient between the mulched strip and the bare strip, water flowing from a mulched strip to a bare strip can partly compensate for water loss in the bare strip. Low water loss in surface and subsurface soil leads to less water transport from deep soil layers (60-100 cm, represented by point 6 in Figure 4-9) to upper soil layers and leads to a more stable SMF in deeper soil profiles.

4.5 Conclusion

ERT monitoring and Hydrus 2D simulation showed the similar behavior of SMF under PM, even though SMF was not identical due to added complexity in the field caused by soil heterogeneity and root distribution. Both ERT monitoring and Hydrus 2D simulation indicated that the application of PM resulted in a contrasting SMF between the mulched strip and the bare strip near the soil surface, and the SMF was more stable in the mulched strip than in the bare strip. The comparison of SMF between PM and NM with Hydrus 2D suggested that the PM stabilizes the SMF for the whole soil section, which means that the whole section experiences fewer fluctuations. The alleviation of SMF is predominant in the mulched strip compared to the bare strip. The inhibition caused by the plastic film and soil water redistribution leads to a higher soil moisture content across the growing season and a narrower moisture fluctuation range in PM. Thus, we conclude that PM is an effective agricultural practice to alleviate strong SMF in rain-fed, semi-arid areas.

4.6 Reference

Allen R. G., Pereira L. S., Raes D., et al, 1998. FAO 56 Irrigation and drainage paper: crop evapotranspiration [M]. Rome: Food and Agriculture Organization.

Beff L., Guenther T., Vandoorne B., et al, 2013. Three-dimensional monitoring of soil water content in a maize field using Electrical Resistivity Tomography [J]. Hydrol. Earth Syst. Sci., 17: 595-609.

Berger S., Kim Y., Kettering J., et al, 2013. Plastic mulching in agriculture-Friend or foe of N2O emissions? [J]. Agric. Ecosyst. Environ., 167: 43-51.

Besson A., Cousin I., Bourennane H., et al, 2010. The spatial and temporal organization of soil water at the field scale as described by electrical resistivity measurements [J]. Eur. J. Soil Sci., 61: 120-132.

Birch H. F, 1958. The effect of soil drying on humus decomposition and nitrogen availability [J]. Plant Soil, 10: 9-31.

Borken W., and Matzner E, 2009. Reappraisal of drying and wetting effects on C and N mineralization and fluxes in soils [J]. Global Change Biol., 15: 808-824.

Chepkwony C. K., Haynes R. J., Swift R. S., et al, 2001. Mineralization of soil organic P induced by drying and rewetting as a source of plant-available P in limed and unlimed samples of an acid soil [J]. Plant Soil, 234: 83-90.

Brunet P., Clement R., and Bouvier C, 2010. Monitoring soil water content and deficit using Electrical Resistivity Tomography (ERT) - a case study in the Cevennes area, France [J]. J. Hydrol., 380: 146-153.

Chen B., Liu E., Mei X., et al, 2018. Modelling soil water dynamic in rain-fed spring maize field with plastic mulching [J]. Agric. Water Manage., 198: 19-27.

Chen B., Yan C., Garré S., et al, 2017. Effects of a "one film for 2 years" system on the grain yield, water use efficiency and cost-benefit balance in dryland spring maize (Zea mays L.) on the Loess Plateau, China [J]. Arch. Agron. Soil Sci., 64: 939-952.

Donat M. G., Lowry A. L., Alexander L. V., et al, 2016. More extreme precipitation in the world's dry and wet regions [J]. Nat. Clim. Change, 6 (5): 508.

FAO, 2006. Guidelines for soil description, 4th edn [M]. Rome: Food and Agriculture Organization.

Feddes R. A., Kowalik P. J., and Zaradny H., 1978. Simulation of field water use and crop yield [M]. New York: John Wiley and Sons.

Feng Y., Liu Q., Tan C., et al, 2014. Water and nutrient conservation effects of different tillage treatments in sloping fields [J]. Arid Land Res. Manage., 28: 14-24.

Gan Y., Siddique K. H. M., Turner N. C., et al, 2013. Chapter seven-ridge-furrow mulching systems-an innovative technique for boosting crop productivity in semiarid rain-fed environments [J]. Adv. Agron., 118: 429-476.

Garre S., Guenther T., Diels J., et al, 2012. Evaluating Experimental design of ERT for soil moisture monitoring in contour hedgerow intercropping systems [J]. Vadose Zone J., 11 (4): 1-14.

Garré S., Javaux M., Vanderborght J., et al, 2011. 3-d electrical resistivity tomography to monitor root zone water dynamics [J]. Vadose Zone J., 10 (1): 412-424.

Ghosh P. K., Dayal D., Bandyopadhyay K. K., et al, 2006. Evaluation of straw and polythene mulch for enhancing productivity of irrigated summer groundnut [J]. Field Crop. Res., 99: 76-86.

Gong D., Mei X., Hao W, et al, 2017. Comparison of ET partitioning and crop coefficients between partial plastic mulched and non-mulched maize fields [J]. Agric. Water Manage., 181: 23-34.

Hu W., Shao M., Han F., et al, 2011. Spatio-temporal variability behavior of land surface soil water content in shrub-and grass-land [J]. Geoderma, 162: 260-272.

Jarvis P., Rey A., Petsikos C., et al, 2007. Drying and wetting of Mediterranean soils stimulates decomposition and carbon dioxide emission: the "Birch effect" [J]. Tree Physiol., 27: 929-940.

Kader M. A., Senge M., Mojid M. A., et al, 2017. Mulching type-induced soil moisture and temperature regimes and water use efficiency of soybean under rain-fed condition in central japan [J]. Int. Soil Water Conserv. Res., 5: 302-308.

Kasirajan S., and Ngouajio M, 2012. Polyethylene and biodegradable mulches for agricultural applications: a review [J]. Agron. Sustain. Dev., 32 (2): 501-529.

Keller G. V., and Frischknecht F. C, 1966. Electrical methods in geophysical prospecting [M]. Oxford: Pergamon Press.

Li R., Hou X., Jia Z., et al, 2013a. Effects on soil temperature, moisture, and maize yield of cultivation with ridge and furrow mulching in the rainfed area of the Loess Plateau, China [J]. Agric. Water Manage., 116: 101-109.

Li S., Kang S., Li F., et al, 2008. Evapotranspiration and crop coefficient of spring maize with plastic mulch using eddy covariance in northwest China [J]. Agric. Water Manage., 95: 1214-1222.

Li S. X., Wang Z. H., Li S. Q., et al, 2013b. Effect of plastic sheet mulch, wheat straw mulch, and maize growth on water loss by evaporation in dryland areas of China [J]. Agric. Water Manage., 116: 39-49.

Li X., Simunek J., Shi H., et al, 2017. Spatial distribution of soil water, soil temperature, and plant roots in a drip-irrigated intercropping field with plastic mulch [J]. Eur. J. Agron., 83: 47-56.

Loke M. H., and Barker R. D, 1996. Rapid least-squares inversion of apparent resistivity pseudosections by a quasi-Newton method [J]. Geophys. Prospect., 44: 131-152.

Lordan J., Pascual M., Villar J. M., et al, 2015. Use of organic mulch to enhance water-use efficiency and peach production under limiting soil conditions in a three-year-old orchard [J]. Span. J. Agric. Res., 13 (4): e0904.

Manzoni S., Schaeffer S. M., Katul G., et al, 2014. A theoretical analysis of microbial eco-physiological and diffusion limitations to carbon cycling in drying soils [J]. Soil Biol. Biochem., 73: 69-83.

NBSC (National Bureau of Statistics of China). 2014. China rural statistical yearbook 2014 [M]. Beijing: China Statistics Press.

Revil A., Wu Y., Karaoulis M., et al, 2013. Geochemical and geophysical responses during the infiltration of fresh water into the contaminated saprolite of the Oak Ridge Integrated Field Research Challenge site, Tennessee [J]. Water Resour. Res., 49 (8): 4952-4970.

Rhoades J. D., Raats P. A. C., and Prather R. J, 1976. Effects of liquid-phase electrical conductivity, water content, and surface conductivity on bulk soil electrical conductivity [J]. Soil Sci. Soc. Am. J., 40 (5): 651-655.

Richards L, 1931. Capillary conduction of liquid in porous media [J]. Physics, 1: 318-333.

Ritchie J. T, 1972. Model for predicting evaporation from a row crop with incomplete cover [J]. Water Resour. Res., 8: 1204-1213.

Samouelian A., Cousin I., Tabbagh A., et al, 2005. Electrical resistivity survey in soil science: a review [J]. Soil Till. Res., 83: 173-193.

Michot D., Benderitter Y., Dorigny A., et al, 2003. Spatial and temporal monitoring of soil water content with an irrigated corn crop cover using surface electrical resistivity tomography [J]. Water Resour. Res., 39 (5): 89-94.

Scarascia-mugnozza G., Sica C., and Russo G, 2011. Plastic materials in European agriculture: actual use and perspectives [J]. J. Agric. Eng., 42: 15-28.

Schimel J., Balser T. C., and Wallenstein M, 2007. Microbial stress-response physiology and its implications for ecosystem function [J]. Ecology, 88: 1386-1394.

Seyfi K., and Rashidi M, 2007. Effect of drip irrigation and plastic mulch on crop yield and yield components of cantaloupe [J]. Int. J. Agric. Biol., 2: 247-249.

Šimůnek J., van Genuchten M. T., and Sejna M, 2008. Development and applications of the HYDRUS and STANMOD software packages and related codes [J]. Vadose Zone J., 7: 587-600.

Solomon S, 2007. Contribution of working group I to the fourth assessment report of the intergovernmental panel on climate change, climate change 2007: the physical science basis [M]. Cambridge: Cambridge University Press.

Turner B. L., and Haygarth P. M, 2001. Biogeochemistry-Phosphorus solubilization in rewetted soils [J]. Nature, 411: 258-258.

UNEP, 1992. World atlas of desertification [M]. London: Edward Arnold.

Vanderborght J., Huisman J. A., Van der Kruk J., et al, 2013. Geophysical methods for field-scale imaging of root zone properties and processes [J]. Soil-Water-Root Processes: Advances in Tomography and Imaging, 247-282.

Vangenuchten M. T, 1980. A closed-form equation for predicting the hydraulic conductivity of unsaturated soils [J]. Soil Sci. Soc. Am. J., 44: 892-898.

Vrugt J. A., Hopmans J. W., and Simunek J, 2001. Calibration of a two-dimensional root water uptake model [J]. Soil Sci. Soc. Am. J., 65: 1027-1037.

Waxman M. H., and Smits L. J. M, 1968. Electrical conductivities in oil-bearing shaly sands [J]. Soc. of Petrol. Eng. J., 8: 107.

Whalley W. R., Binley A., Watts C. W., et al, 2017. Methods to estimate changes in soil water for phenotyping root activity in the field [J]. Plant Soil, 415 (1-2): 407-422.

Zhang Y., Xiao Q., and Huang M, 2016. Temporal stability analysis identifies soil water relations under different land use types in an oasis agroforestry ecosystem [J]. Geoderma, 271: 150-160.

Zhao H., Wang R. Y., Ma B. L., et al. Ridge-furrow with full plastic film mulching improves water use efficiency and tuber yields of potato in a semiarid rainfed ecosystem [J]. Field Crop. Res., 161: 137-148.

Zhao W., Cui Z., Zhang J., et al, 2017. Temporal stability and variability of soil-water content in a gravel mulched field in northwestern China [J]. J. Hydrol., 552: 249-257.

5

Effects of Plastic Mulching on Soil Microbial Community

5.1 Abstract

Information about the effect of plastic film mulching (PFM) on the soil microbial communities of rain-fed regions remains scarce. In the present study, Illumina Hiseq sequencer was employed to compare the soil bacterial and fungal communities under three treatments: no mulching (NM), spring mulching (SM) and autumn mulching (AM) in two layers (0-10 and, 10-20 cm). Our results demonstrated that the plastic film mulching (PFM) application had positive effects on soil physicochemical properties as compared to no-mulching (NM): higher soil temperature (ST), greater soil moisture content (SMC) and better soil nutrients. Moreover, mulching application (especially AM) caused significant increase of bacterial and fungal richness and diversity and played important roles in shaping microbial community composition. These effects were mainly explained by the ST and SMC induced by the PFM application. The positive effects of AM and SM on species abundances were very similar, while the AM harbored relatively more beneficial microbial taxa than the SM, e.g., taxa related to higher degrading capacity and nutrient cycling. According to the overall effects of AM application on ST, SMC, soil nutrients and microbial diversity, AM is recommended during maize cultivation in rain-fed region of northeast China.

Keywords: Soil Microbial Communities; Plastic Mulching; Rain-fed Agriculture; Maize.

5.2 Introduction

Plastic film mulching (PFM) has been used for crop production worldwide since the 1950s, and its use has increased, particularly in arid and semi-arid areas (Wang et al., 2016). Over 30 million acres of agricultural land worldwide were covered with plastic mulch as of 1999, and an estimated 1 million tons of mulch films are used annually in the agricultural sector (Cuello et al., 2015). Maize is the most important field crop in cold and arid regions of northeast China and occupies the largest cultivated area. Low temperature and water shortage are the major limiting factors for maize production (Bu et al., 2013). The application of film mulching technology has expanded the maize planting area and enhanced maize production by 10-15 billion kilograms per year, which is the equivalent of 5-8% of the total national maize output in China (Liu et al., 2014).

The soil microbial community plays a crucial role in nutrient cycling, maintenance of soil structure and its diversity is a sensitive indicator of soil quality that can reflect subtle changes and provide information for evaluation of soil function (Larkin et al., 2003). The determination of factors that influence microbial community composition under field conditions has significantly increased our understanding of how management affects crop quality, disease ecology, and biogeochemical cycling (Maul et al., 2014). Thus, research on the effects of soil management on soil microbial communities has become a fundamental aspect of sustainable agriculture.

Studies have confirmed that PFM is generally associated with physical and chemical changes in soil characteristics, such as decreases in the amount of water loss caused by evaporation (Kasirajan et al., 2012), enhancement of soiltemperatures (Wang et al., 2015) and changes in soil nutrient availability (Wang et al., 2016). Previous studies mainly focused on the impacts of PFM on soil moisture, soil structure, soil nutrition and crop yield, while their influences on soil micro-ecological environment and evolution of soil microbial community were neglected (Steenwerth et al., 2002). Just as different groups of microorganisms vary in their ability to adapt to the various soil environmental conditions, changes in soil microbiology under PFM application are expectable (Muñoz et al., 2015). Li et al. (2004) demonstrated that PFM application increased the soil microbial activity in a spring wheat field, but the extent of this effect depended on how long the mulching was maintained during the growing season. In a study conducted in a maize cropping system, PFM showed higher microbial biomass carbon (C) and nitrogen (N) and several enzyme activities of C, N, and phosphorus (P) cycling compared with no mulching (NM) (Wang et al., 2014). However, the detailed effects of continuous PFM on soil microbial communities and the link between these effects and the soil environment remain unclear. Particularly, Chen et al. (2014) indicated that bacterial communities in soils that underwent various mulching treatments obviously differed from those in soils that received NM in an orchard, and they were associated with different soil physicochemical conditions. Liu et al. (2012) found that PFM treatment could significantly change the community composition of arbuscular mycorrhizal fungi under a spring wheat cropping system in a temperate semi-arid region. Soil fungi and bacteria exhibit different community dynamics (Prewitt et al., 2014); thus, to understand soil ecosystem processes, it is essential to address fungal and bacterial communities simultaneously (Baldrian et al., 2012). However, how both bacterial and fungal communities respond to the microclimate conditions generated by PFM has not been elucidated together in a single study so far.

The western area of Liaoning Province is a typical semi-arid rain-fed agricultural region of China in which the limited rainfall cannot meet the agricultural production demands. Previous research has demonstrated that PFM treatments, particularly autumn mulching (AM) and spring mulching (SM), were the most optimal technical models for increasing water use efficiency and maize yield in these areas (Zou et al., 2005). In the local agricultural production, SM application was conducted in spring of the present year until the harvest, while AM application was conducted in autumn of the previous year until the harvest. Therefore, the length of film mulching time under AM and SM were fixed and different. However, the effects of different mulching treatments on the microbial community structure have not been assessed in a comparative manner. Since PFM application generates artificial microclimatic conditions which may have an influence on the microbial diversity and abundance in soil, the main objective of the present work was to evaluate the impacts of different PFM approaches on the soil microbial community during rainfed maize production. With this aim, the phylum to genus levels of the composition and diversity of bacterial and fungal communities under the soil of two mulching treatments and NM were compared. Moreover, the relationships of the microbial community structure and diversity with the physicochemical properties of the PFM soils were analysed. We hypothesized that ①both the AM and SM mulching treatments would lead to a higher soil microbial diversity and a restructuring of soil microbial communities compared with no-mulching treatment, and that ②the variation in bacterial and fungal communities might also be associated with soil environmental conditions, e. g. soil organic carbon (SOC) and soil moisture content (SMC). To our knowledge, this is the first study to investigate bacterial and fungal communities under different PFM treatments.

5.3 Materials and methods

5.3.1 Site description

Field experiments were conducted at the Fuxin Agricultural Environment and Farmland Conservation Experimental Station of China's Ministry of Agriculture in Fuxin County (42°09′N, 121°46′E), Liaoning Province, and northeast China, which has a semi-arid climate. From 2011 to 2014, the mean annual air temperature was 7.2℃, the average annual precipitation was 490 mm, and 73% of this precipitation fell during the maize growing season (from April to September). The average potential evaporation was 1,830 mm. The soil at the experiment site is classified as Udalfs according to the USDA soil taxonomy system. Based on the investigation and analysis before our experiment, the soil properties at the top 20 cm were as follows: bulk density, 1.35 g·cm^{-3}; pH, 6.95; soil organic matter, 6.1 g·kg^{-1}, total nitrogen (TN), 0.456 g·kg^{-1}; total phosphorus (P), 0.66 g·kg^{-1}; total potassium (K), 2.46 g·kg^{-1}; available N, 196.6 mg·kg^{-1}; available P, 136.1 mg·kg^{-1} and available K, 62.2 mg·kg^{-1}.

5.3.2 Field experiment design

In 2011, the experiment was composed of three different PFM treatments: ①no mulching (NM); ②spring mulching (SM): mulching in spring of the present year (from April 8th to the harvest); ③autumn mulching (AM), mulching in autumn of the previous year (from November 10th to the harvest). Each treatment plot (10 m long×5 m wide) was replicated three times and laid out in a randomized complete block design. In PFM treatments, the whole land surface (all ridges and furrows) was covered with polyethylene film (colorless and transparent, each strip was 0.008 mm thick and 1.2 m wide) for the soil, and the amount of film was approximately 75 kg·ha^{-1}.

At the end of April, a maize cultivar "Zhengdan 958" was sown at the plant density of

60,000 plants·ha^{-1} using a hole-sowing tool. All plots were harvested at the end of September. In 2014, the maize yields in different treatments were as follows: NM (12,470 ± 363.059 kg·ha^{-1}), SM (12,024 ± 312.692 kg·ha^{-1}) and AM (14,280 ± 330.327 kg·ha^{-1}). The harvest was followed by a fallow period until sowing time. Each plot was supplied with N 240 kg·ha^{-1} (urea, N, 46%), P 140 kg·ha^{-1} (calcium superphosphate, P_2O_5, 12%) and K 190 kg·ha^{-1} (potassium sulfate, K_2SO_4, 45%) at sowing.

5.3.3 Sampling collection and analysis

After four years of continuous mulching treatments, soil samples were collected from the plots at the time of harveston September 18th, 2014. In the PFM plots, the soil cores were sampled by penetrating the plastic film. The mulching materials on the sampling points were carefully removed during the collection of soil samples (Li et al., 2015). For all plots, soil at five randomly selected locations was sampled in two layers (0-10 cm and 10-20 cm, referred to as the surface and sub-surface, respectively) using an auger with a 5 cm internal diameter and then mixed as one sample. Every year after harvest, the plastic films were gathered and recycled. All of the fresh soil samples were air-dried and sieved twice, using 2.0 mm and 0.25 mm meshes. Each sample was then divided into two portions: one portion was stored at 4℃ for chemical analysis, and the other portion was stored at -80℃ for DNA extraction.

5.3.3.1 Soil chemical property analysis

The soil chemical properties were measured using the methods described by Bao19. Soil moisture content (SMC) and soil temperature (ST) were monitored in the fields using EcH_2O (Decagon Devices, WA, USA) soil moisture and temperature sensors which were attached to a data logger from sowing to harvest, Soil bulk density (BD) was determined using the core method. The SOC level was measured with a K_2CrO_7-H_2SO_4 oxidation method. The TN was measured using the Kjeldahl digestion method. The dissolved organic carbon (DOC) content of the filtered 0.5 M K_2SO_4 extracts from the fresh soil. Concentrations of DOC and total dissolved nitrogen (TDN) were determined with a TOC analyser (Liqui TOC Elementar, Vario Max, Germany). Concentration of dissolved organic nitrogen (DON) was calculated as the difference between the TDN reading and the combined ammonium-N and nitrate-N source reading.

5.3.3.2 DNA extraction and PCR amplification

Microbial DNA was extracted from 0.5 g soil samples using a Power Soil DNA Kit (MOBIO Inc., Carlsbad, CA) according to the manufacturer's protocols. PCR was performed with the Takara ExTaq PCR kit (Takara Shuzo, Osaka, Japan). The V3-V4 regions of the bacterial 16S rRNA gene were amplified using the following PCR conditions: 95℃ for 3 min; followed by 30 cycles of 95℃ for 30 s, 55℃ for 30 s, and 72℃ for 45 s; and a final extension at 72℃ for 10 min. The following PCR primers were used for this amplification: 338F 5'-ACTCCTACGGGAGGCAGCA-3' and 806R 5'-GGACTACHVGGGTWTCTAAT-3'. The region of the fungal ITS ribosomal gene was amplified using the following PCR conditions: 95℃ for 3 min, followed by 34 cycles at 95℃ for 30 s, 55℃ for 30 s, and 72℃ for 45 s and a final extension at 72℃ for 10 min. The primers ITS1F 5'-CTTGGTCATTTAGAGGAAGTA A-3' and ITS2 5'-GCTGCGTTCTTCATCGATGC-3' were used (Huttenhower et al., 2013).

5.3.3.3 Illumina HiSeq sequencer

The amplicons were extracted from 2% agarose gels and purified using the AxyPrep DNA Gel Extraction Kit (Axygen Biosciences, Union City, CA, USA.) and quantified using Quanti-Fluor™-ST (Promega, USA.) according to the manufacturer's instructions. Each sample was quantified using a Qubit 3.0 Fluorometer (Life Technologies, Grand Island, NY), then pooled at equal concentrations and diluted to create one sample. Paired-end with a read length of 2×250 bp was carried out on an Illumina HiSeq platform according to standard protocols. Raw FASTQ files were de-multiplexed and, quality-filtered using QIIME 1.17 with the following criteria: ①exact primer matching, two nucleotide mismatches in primer matching, and reads containing ambiguous characters were removed; and ②initial quality control measures removed any sequence with a consensus fold-coverage <5, average quality score <25 (50 bp rolling window). The reads that could not be assembled were discarded. All sequences with ambiguous base calls were discarded (bacteria: 933,972 reads in total and 825,614 effective reads, fungi: 1,117,315 reads in total and 1,088,110 effective reads). Effective sequences were normalized for the following analysis. The raw readings were deposited into the NCBI Sequence Read Archive (SRA) database with bioproject accession number PRJNA350374: SAMN05949267 to SAMN05949284 for bacteria and SAMN05949285 to SAMN05949302 for fungi.

5.3.3.4 Statistical analysis

All of the results were reported as the means and standard error (SE). Our data passed three assumptions before analysis: ①there were no significant outliers; ②the dependent variable was normally distributed for each combination of the groups of the two independent variables; ③our data variance was homogeneous for each combination of the groups of the two independent variables. The effects of mulching treatments, soil layers, and their combined interaction on fungal and bacterial assemblage diversity and richness and soil properties were determined by two-way analysis of variance (ANOVA) (SPSS Statistics, Version 18 IBM Corp., Armonk, NY, USA), and statistically significant differences were tested using Duncan's multiple range test. Data analysis followed the steps below: Test interaction to see if there are combination effects. If so, see which combinations are different from the others and by how much, if no interactions, test the factors separately. If a factor is important (large F), decide which of its means are different and by how much. In every soil layer, one-way analysis of variance (ANOVA) and Duncan's multiple comparisons were performed to identify differences in the microbial relative abundances among the three treatments. A difference at $P<0.05$ was considered statistically significant (Table 5-1).

Table 5-1 Coverage value for different soil samples across all soil treatments

Sample ID	Coverage value	
	Bacterial	Fungal
SM0-10-1	0.93069	0.98861
SM0-10-2	0.95837	0.97520
SM0-10-3	0.96812	0.99652
AM0-10-1	0.97069	0.99357
AM0-10-2	0.96346	0.99232
AM0-10-3	0.96438	0.99716
NM0-10-1	0.91228	0.99236
NM0-10-2	0.93822	0.99443
NM0-10-3	0.93795	0.99412
SM10-20-1	0.96188	0.99738
SM10-20-2	0.95084	0.99639
SM10-20-3	0.85402	0.98079
AM10-20-1	0.88996	0.98723
AM10-20-2	0.93912	0.99182
AM10-20-3	0.93167	0.99190
NM10-20-1	0.95376	0.99707
NM10-20-2	0.97034	0.99631
NM10-20-3	0.93705	0.99266

Operational taxonomic units (OTUs) were clustered with a 97% similarity cut off using Uclust version v1.2.22q, and chimeric sequences were identified and removed using UCHIME. Completeness of the sampling effort was evaluated using Good's rarefaction curves and coverage. The phylogenetic affiliation of each 16S rRNA gene sequence was analysed with RDP Classifier (http://rdp.cme.msu.edu) against the SILVA (SSU123) 16S rRNA database using a confidence threshold of 60%. The α-diversity analysis, which was based on Mothur v.1.30.1, was conducted to reveal the Chao, Shannon and Simpson diversity indices. In species richness estimation, the Chao1 index is the most commonly used, and which is based upon the number of rare classes (i.e., OTUs) found in a sample. In an ecosystem, the higher the Chao1 index presented, the richer the system. The Shannon index and the Simpson index were defined as being the most sensitive to changes in the importance of the species in the sample. Hence, to characterize species diversity in a community, we adopted two of the most widely used indices (Shannon index and Simpson's index). The Shannon index measured the average degree of uncertainty in predicting the species of an individual chosen at random from a collection, and the value increases as the number of species increases and as the distribution of individuals among the species becomes even. The Simpson's index indicates species dominance and reflects the probability of randomly choosing two individuals that belong to the same species. It varies from 0 to 1, and the index increases as the diversity decreases. For ß-diversity analysis, Principal Coordinates Analysis (PCoA) was performed to assess the similarities of the samples' community memberships. The goal of PCoA is to decrease the dimensionality while preserving the pairwise dissimilarity values as much as possible in the data set (Legendre and Legendre, 2012).

In our analysis, the weighted UniFrac distance metrics (based on phylogenetic structure) were used to generate PCoA using the principal_coordinates. py command in QIIME (Lozupone et al., 2011). To determine the key factor (s) affecting microbial diversity, stepwise multiple regression analysis was applied using the probability criteria of $P<0.05$ to accept and $P>0.1$ to remove a variable from the analysis. Direct ordination was either by canonical correspondence analysis (CCA) or redundancy discriminate analysis (RDA), depending on the length of the detrended correspondence analysis (DCA) axis (where an axis of $>4.0=$ CCA and an axis of $<4.0=$ RDA, "decorana" function in vegan). The initial DCA results demonstrated that the data exhibited unimodal rather than linear responses to the environmental variables. Therefore, CCA was carried out using the "cca" function in the vegan package of R (version 2.15.1) to assess the relationships between environmental parameters and microbial community structures (Xie et al., 2015). The significances of environmental factors (included, SMC, ST, BD, SOC, DOC, TN and DON) were assessed using the "envfit" function, which after determining r2 for environmental variables uses a permutation procedure to define the significance of each environmental variable (999 permutations) on all axes conjointly. Variables with no significant differences were removed from the analysis.

5.4 Results

5.4.1 *Soil physical and chemical parameters*

The SMC was always higher in the mulched than in the NM plots, and the differences were more obvious in middle growing season (Figure 5-1). For different treatments, the AM plots had higher SMC than that under the other two treatments for most sampling dates during the whole growing stage. The average SMC under the AM treatment in surface and subsurface soil was 4%, and 6% higher than those in the SM and NM treatments, respectively. Similarly, the ST in the soils was slightly higher in the PFM treatments than those in the no-mulching treatments during most of the growing periods. The ST ranged from 17.6 to 28.9℃, with an average of 22.7℃ under AM in surface soil, and from 17.4 to 26.5℃, with an average of 21.7℃ in subsurface. Throughout the growing season, the ST in the AM was, on average, 1.5℃ and 2.0℃ higher than those in the SM and NM, respectively.

For the investigated soil properties after the harvesting stage the average TN, DOC and DON contents in AM were highest and were approximately 58%, 39% and 101% higher than those of the NM treatment, respectively (Table 5-2). However, compared with the NM, the mulching treatments had no significant influence on SOC and BD contents ($P>0.05$). The SMC, ST, BD, SOC and DOC were found to be significantly different between two layers ($P<0.05$). Specifically, the SMC and BD at surface were significantly higher than that in subsurface. In contrast, the SOC and DOC surface were obviously less than that in subsurface. Moreover, there were no significant differences were observed in TN and DON between two layers.

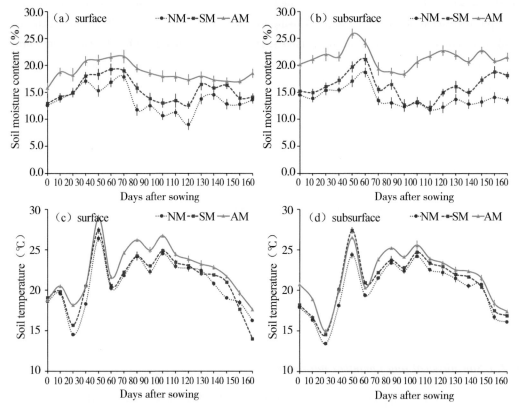

Figure 5-1 Changes in soil temperatures in surface (a) and subsurface (b) and moisture contents in surface (c) and subsurface (d) under different mulching treatments during maize cultivation

5.4.2 Richness and diversity of bacterial and fungal communities

Bacterial and fungal richness and diversity are shown in Table 5-3. The ANOVA results showed that both the mulching treatment and soil layer affected bacterial and fungal community richness (Chao1) and the highest Chao1 values for both bacterial and fungal communities were observed under AM treatment. For diversity indices, bacterial and fungal communities displayed different trends. Specifically, the Shannon index of bacterial communities in soils with mulching (AM:

7.31 and SM: 7.16, respectively) were significantly higher than those in soils with NM (6.75), and the Simpson indices (AM: 0.002 and SM: 0.003, respectively) were clearly lower than those of soils with NM (0.007). In addition, the soil layer effect on the bacterial diversity indices was not significant. In contrast, both the mulching treatments and the soil layer significantly affected fungal richness and diversity ($P<0.05$), and AM treatment soil showed the greatest values of richness and the Shannon index, as well as the lowest Simpson index.

Table 5-2 Effects of different mulching treatments on the physical and chemical properties of the soils sampled at two layers

layer	Treatment	SMC (%)	ST (℃)	BD (g·cm^{-3})	SOC (g·kg^{-1})	TN (g·kg^{-1})	DOC (mg·kg^{-1})	DON (mg·kg^{-1})
Surface	NM	14 (0.6)	20.1 (0.1)	1.28 (0.04)	6.3 (0.0)	0.58 (0.01)	41.7 (4.9)	82.2 (6.5)
	SM	17 (0.2)	20.9 (0.1)	1.27 (0.04)	6.2 (0.1)	0.49 (0.07)	48.7 (3.9)	126.7 (29.8)
	AM	19 (0.2)	21.4 (0.1)	1.28 (0.05)	6.5 (0.0)	0.63 (0.11)	56.9 (2.1)	227.6 (31.3)
Subsurface	NM	25 (0.1)	19.5 (0.1)	1.48 (0.02)	4.9 (0.2)	0.41 (0.09)	34.4 (3.7)	102.4 (6.2)
	SM	26 (0.9)	19.8 (0.1)	1.44 (0.03)	5.1 (0.3)	0.30 (0.06)	43.9 (1.4)	124.3 (47.8)
	AM	28 (1.0)	20.3 (0.1)	1.38 (0.03)	5.3 (0.1)	0.63 (0.11)	48.6 (3.4)	141.7 (27.1)
Mulching effect (M)								
	NM	19 c	19.8 c	1.38 ns	5.6 ns	0.40 c	38.1 b	92.3 b
	SM	21 b	20.3 b	1.36 ns	5.6 ns	0.50 bc	46.4 b	125.5 ab
	AM	24 a	20.9 a	1.33 ns	5.9 ns	0.63 a	52.8 a	184.6 a
Layer effect (D)								
	surface	17 b	20.8 a	1.28 b	6.3 a	0.57ns	49.1 a	145.5 ns
	subsurface	26 a	19.9 b	1.44 a	5.1 b	0.45ns	42.3 b	122.8 ns
Significance								
	Mulching (M)	**	**	ns	ns	*	**	*
	Layer (L)	**	**	**	**	ns	*	ns
	M×L	**	ns	ns	ns	ns	ns	ns

Values are means (with standard error in parentheses). The values in mulching effect or layer effect were given the mean under each mulching treatment or soil layer. SMC: soil moisture content; ST: soil temperature; BD: Bulk density; SOC: soil organic carbon; TN, total nitrogen; DOC, dissolved organic carbon; DON: dissolved organic nitrogen; Values within columns followed by the same letter do not differ at< 0.05. ns: not significant. * Significant at 0.05 level. ** Significant at 0.01 level.

Table 5-3 Effects of different mulching treatments on bacterial and fungal α-diversity

Layer	Treatment	Bacterial			Fungal		
		Chao1	Simpson	Shannon	Chao1	Simpson	Shannon
Surface	SM	7153 (58)	0.003 (0.000)	7.16 (0.00)	1393 (57)	0.0678 (0.008)	4.00 (0.02)
	AM	7433 (140)	0.002 (0.000)	7.35 (0.01)	1610 (101)	0.049 (0.002)	4.10 (0.04)
	NM	6126 (57)	0.007 (0.001)	6.87 (0.13)	1139 (31)	0.099 (0.005)	3.59 (0.03)

(continued)

Layer	Treatment	Bacterial			Fungal		
		Chao1	Simpson	Shannon	Chao1	Simpson	Shannon
Subsurface	SM	6406 (51)	0.003 (0.000)	7.17 (0.03)	1240 (23)	0.082 (0.004)	3.78 (0.08)
	AM	6707 (82)	0.002 (0.001)	7.28 (0.02)	1343 (40)	0.069 (0.002)	3.93 (0.06)
	NM	5366 (131)	0.008 (0.000)	6.64 (0.22)	751 (91)	0.168 (0.010)	3.18 (0.02)
Mulching effect (M)							
	SM	6780 b	0.003b	7.16 a	1317 b	0.059 c	3.38 c
	AM	7070 a	0.002b	7.31 a	1477 a	0.074 b	4.01 a
	NM	5746 c	0.007a	6.75 b	946 c	0.133 a	3.89 b
Layer effect (L)							
	Surface	6904 b	0.004 ns	7.13 ns	1380 a	0.071 b	3.89 a
	Subsurface	6159 a	0.004 ns	7.03 ns	1111 b	0.106 a	3.63 b
Significance							
	Mulching (M)	**	**	**	**	**	**
	Layer (L)	**	ns	ns	**	**	**
	M×L	ns	ns	ns	ns	**	ns

Values are shown as the means (with standard error in parentheses). The values in mulching effect or layer effect were given the means under each mulching treatment or soil layer. $*P<0.05$; $**P<0.01$; ns, not significant.

5.4.3 Treatment effects on bacterial and fungal β-diversity

UniFrac-weighted PCoA based on the OTU composition also clearly demonstrated variations among the different soil samples, with the first two axes explaining 52.2% and 23.6% of the

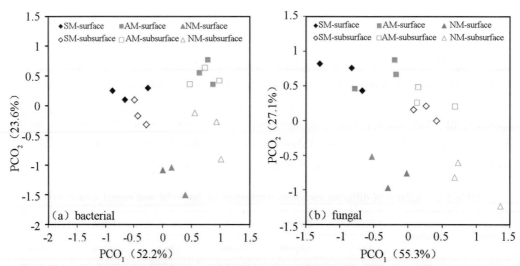

Figure 5-2　UniFrac-weighted PCoA plots of bacterial (a) and fungal (b) communities for different treatments in the two layers

total variation for the bacterial and 55.3% and 27.1% for the fungi (Figure 5-2), respectively. For bacteria, the NM treatment was distinctly separated from SM and AM treatments along the first component (PCoA1) in surface soil, while the AM treatment was separated from the SM and NM treatments along the second component (PCoA2) in the subsurface. For fungi, the NM treatment was clearly separated from the SM and AM treatments along the second component (PCoA2) in both layers. For the different soil layers, fungi exhibited more obvious separation than bacteria along the first component.

5.4.4 *Predominant bacterial and fungal taxa*

Taxonomic composition of bacterial and fungal communities at the phylum level. The phylum level composition of the bacterial and fungal communities varied among the different mulching treatments (Figure 5-3). The bacterial representatives of the phyla *Proteobacteria* and *Acti-*

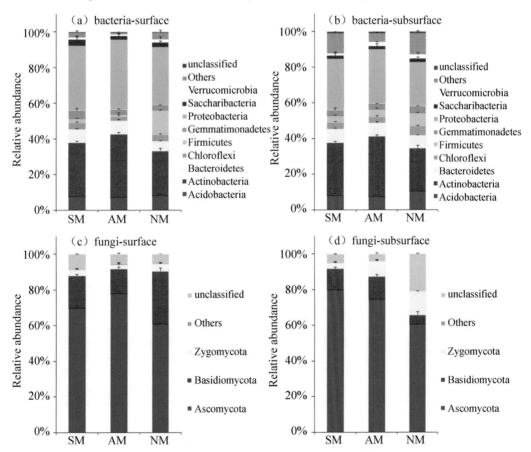

Figure 5-3 Taxonomic distributions of bacterial phyla in surface (a) and subsurface (b) soil samples and of fungal phyla in surface (c), and, subsurface (d) soil samples. Phyla having a mean relative abundance of at least 1% are shown individually, and less abundant phyla are grouped into other

nobacteria were dominant for all treatments in surface and subsurface soil communities. Comparing the differences in the abundance of these phyla, in surface soil, mulching treatment (especially the AM treatment) revealed a significantly higher abundance of sequences affiliated with *Actinobacteria*, *Bacteroidetes* and *Proteobacteria*, while samples from the NM treatment had significantly higher numbers of sequences related to *Chloroflexi* and *Firmicutes* ($P<0.05$) (Table 5-4). In subsurface soil, the mulching treatment samples had greater abundances of sequences affiliated with *Actinobacteria*, and *Proteobacteria* than the NM treatment samples. Similarly, the NM treatment samples had higher numbers of sequences related to *Chloroflexi* and *Firmicutes*. No pronounced differences in the relative abundances of the other four phyla (*Acidobacteria*, *Gemmatimonadetes*, *Saccharibacteria* and *Verrucomicrobia*) were observed between the communities for either soil layer.

Table 5-4 Average relative abundances of bacterial phyla in surface (a) and, subsurface (b) soil samples and of fungal phyla in surface (c) and, subsurface (d) soil samples across all soil treatments

Layer		Surface			Subsurface		
Treatments		SM	AM	NM	SM	AM	NM
Bacteria	Acidobacteria	7.70 (0.012) ns	7.06 (0.003) ns	8.23 (0.004) ns	7.72 (0.010) ns	7.41 (0.007) ns	10.31 (0.009) ns
	Actinobacteria	30.22 (0.007) b	35.64 (0.009) a	24.96 (0.016) c	29.92 (0.006) ab	33.69 (0.024) a	24.22 (0.016) b
	Bacteroidetes	7.43 (0.004) b	7.49 (0.008) a	5.49 (0.002) c	7.85 (0.011) ns	7.54 (0.013) ns	7.33 (0.004) ns
	Chloroflexi	2.63 (0.002) b	2.27 (0.005) b	3.75 (0.004) a	3.60 (0.006) b	3.33 (0.003) b	5.18 (0.002) a
	Firmicutes	2.15 (0.005) b	1.40 (0.002) b	13.42 (0.004) a	3.63 (0.003) b	4.36 (0.003) b	7.12 (0.002) a
	Gemmatimonadetes	4.73 (0.012) ns	2.62 (0.001) ns	3.08 (0.001) ns	3.13 (0.003) ns	3.13 (0.003) ns	3.93 (0.006) ns
	Proteobacteria	36.53 (0.012) ab	39.35 (0.012) a	32.70 (0.006) b	29.04 (0.004) a	30.84 (0.003) a	24.75 (0.002) b
	Saccharibacteria	3.34 (0.004) ns	1.98 (0.003) ns	2.53 (0.003) ns	1.69 (0.014) ns	1.68 (0.007) ns	2.15 (0.006) ns
	Verrucomicrobia	1.24 (0.002) ns	0.76 (0.001) ns	1.65 (0.001) ns	1.11 (0.004) ns	2.09 (0.003) ns	2.03 (0.003) ns
	Others	2.80 (0.007)	1.12 (0.002)	3.94 (0.012)	11.40 (0.002)	5.03 (0.004)	12.10 (0.006)
	unclassified	0.23 (0.001)	0.30 (0.001)	0.24 (0.001)	0.89 (0.001)	0.90 (0.001)	0.87 (0.002)
Fungi	Ascomycota	69.64 (0.033) a	78.03 (0.046) a	60.63 (0.031) b	79.91 (0.022) a	74.60 (0.041) a	60.29 (0.134) b
	Basidiomycota	18.31 (0.066) ab	13.88 (0.036) b	29.95 (0.012) a	11.87 (0.013) a	12.78 (0.009) a	5.53 (0.003) b
	Zygomycota	3.46 (0.010) ns	2.12 (0.003) ns	4.19 (0.021) ns	3.20 (0.008) ns	8.38 (0.036) ns	13.13 (0.075) ns
	Others	0.20 (0.002)	0.58 (0.002)	0.27 (0.001)	0.20 (0.001)	0.11 (0.001)	0.10 (0.001)
	unclassified	8.39 (0.042)	5.38 (0.012)	4.96 (0.008)	4.82 (0.006)	4.13 (0.001)	20.94 (0.006)

The different letter in the same line indicates significant difference at $P<0.05$ among three treatments.

Across all treatments in the two soil layers, the three most dominant fungal phyla were *Ascomycota*, *Basidiomycota* and *Zygomycota*, which accounted for more than 90% of the total fungal sequences (Figure 5-3). For fungi, *Ascomycota* were significantly more abundant in mulching samples (AM and SM) than in the NM samples for both the layers ($P<0.05$). Nevertheless, the abundance of *Basidiomycota* in the NM samples was significantly higher than that in the AM samples in surface soil and lower in subsurface soil ($P<0.05$).

Taxonomic composition of bacterial and fungal communities were analyzed at the class level. Patterns of taxonomic distribution became more apparent at the class levels. The five most abundant bacterial classes found in mulching treatment soil samples were *Actinobacteria*, *Alphaproteobacteria*, *Acidobacteria*, *Gammaproteobacteria*, and *Sphingobacteria* (Figure 5-4). In the surface soil, the AM treatment samples had significantly greater abundances of sequences affiliated with *Actinobacteria* than other two treatments ($P<0.05$). In the subsurface soil samples, the AM treatment samples had greater abundances of sequences affiliated with *Actinobacteria*, *Alphaproteobacteria*, and *Betaproteobacteria* than the other two treatment samples. Meanwhile, the SM treatment samples had higher numbers of sequences related to *Thermomicrobia* than other two treatments samples in the subsurface. For fungi, classes affiliated with *Eurotiomycetes*, *Dothideomycetes* and *Sordariomycetes* were the most abundant, consisting of greater than 70% of the classified fungal classes for both soil layers. The abundances of *Eurotiomycetes* were significantly higher in the mulching treatment samples than in the NM treatment samples ($P<0.05$), whereas the abundances of *Dothideomycetes* were higher in the NM treatment samples than in the mulching treatment samples.

Taxonomic composition of bacterial and fungal communities were also analyzed at the genus level. Based on relative abundance, five dominant bacterial genera were *Agrobacterium*, *Pseudomonas*, *Corynebacterium*, *Streptomyces* and *Lysobacter* (Figure 5-5). The relative abundances of *Streptosporangium* and *Cytophaga* in surface and subsurface soil samples had different change trends: *Streptosporangium* was more enriched in the surface soil samples, whereas *Cytophaga* was more abundant in the subsurface soil samples. Microorganisms from the bacterial genera *Corynebacterium* (*Actinobacteria*), *Frankia* (*Actinobacteria*), *Nitrobacter* (*Proteobacteria*), *Cellulomonas* (*Actinobacteria*), *Streptomyces* (*Actinobacteria*), *Agrobacterium* (*Proteobacteria*) and *Pseudomonas* (*Proteobacteria*) showed higher abundances in the AM and SM treatments soil samples (particularly AM) than that in the NM treatment soil samples. However, the relative abundances of *Lysobacter* and *Stenotrophomonas* were significantly higher in the NM treatment soil samples. Distinctions in the composition of fungal communities were also observed at the genus level (Figure 5-6). The most abundant genera found in both layers were *Penicillium* (*Ascomycota*), *Chaetomium* (*Ascomycota*) and *Trichoderma* (*Ascomycota*). Comparing the differences in the proportion of all abundant genera, AM treatment soil samples had greater abundances of sequences affiliated with *Penicillium*, *Talaromyces*, *Trichoderma* and *Cryptococcus*, while SM treatment soil samples had higher numbers of se-

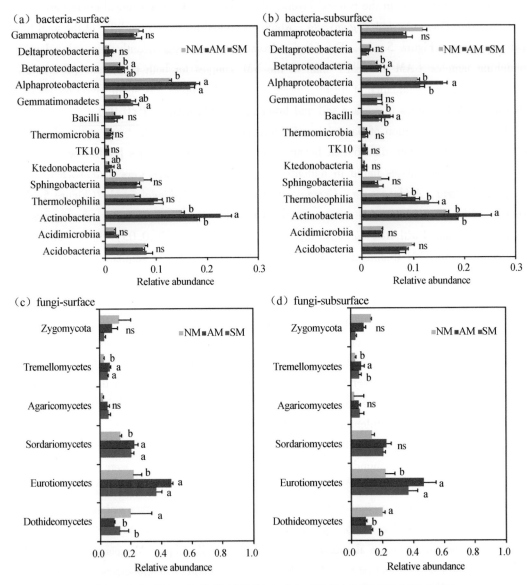

Figure 5-4 Taxonomic distributions of predominant classes of bacteria (a, b) and of fungal classes in surface (c), and, subsurface (d) soil samples. The different letter indicates significant difference at $P<0.05$

quences related to *Fusarium* than NM treatment soil samples. In contrast, NM treatment obviously enriched the abundance of *Cladosporium*. Moreover, the major genera were more abundant in surface soil relative to subsurface soil samples, except *Mortierella*, which presented higher abundances in subsurface soil samples.

A stepwise regression analysis was conducted to determine the correlation of soil environmental parameters with microbial community α-diversity (Table 5-5). The results revealed that ST significantly affected bacterial and fungal α-diversity index, including bacterial Chao1,

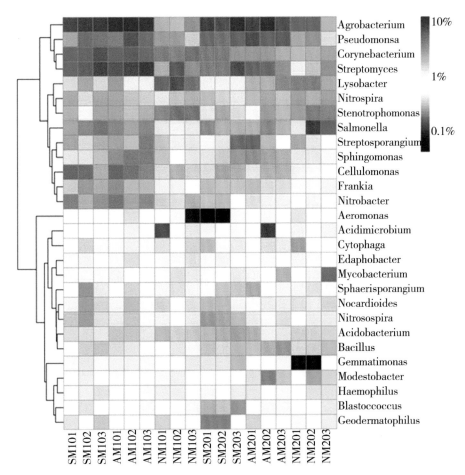

Figure 5-5 Heat map showing the abundance of bacterial genera in each sample (the genera with an average abundance greater than 0.5% in one group were defined as abundant). The colour intensity (log scale) in each panel shows the percentage of a genus in a sample; please refer to the colour key at the right bottom

Simpson index and Shannon index, fungal Chao1, Simpson index and Shannon index. The SOC was also an important factor which clearly influenced bacterial Simpson index and fungal Chao1.

Table 5-5 The variables which were found by stepwise regression analysis to be correlated with bacterial and fungal α-diversity index

	Dependents	Variables related	R^2	Significance
	Chao1	ST	0.771	< 0.001
Bacterial	Simpson	ST SOC	0.760	< 0.001
	Shannon	ST	0.501	< 0.001

	Dependents	Variables related	R^2	Significance
	Chao1	ST SOC	0.794	< 0.001
Fungal	Simpson	ST	0.677	< 0.001
	Shannon	ST	0.666	< 0.001

(continued)

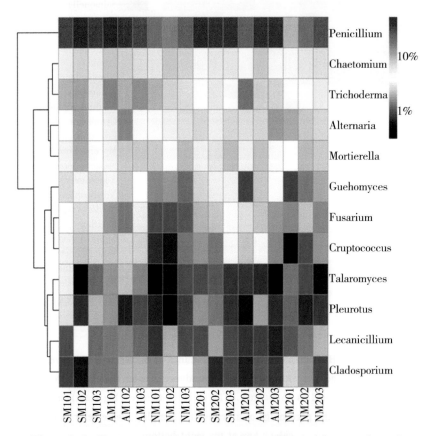

Figure 5-6 Heat map showing the abundance of bacterial genera in each sample (the genera with an average abundance greater than 0.5% in one group were defined as abundant). The colour intensity (log scale) in each panel shows the percentage of a genus in a sample; please refer to the colour key at the right bottom

To further explore the correlation of soil environmental parameters with community structures of bacteria and fungi, we performed a canonical correlation analysis (CCA) based on the sequencing data (Figure 5-7). Mantel test analyses indicated that SMC, SOC and ST were significantly correlated with soil bacterial communities. SMC (F = 5.016, P = 0.003) was the most significantly correlated variable with soil bacterial communities at the genus level, followed by the SOC (F = 4.789, P = 0.006) and ST (F = 4.633, P = 0.008). In addition, we also investigated the soil factors that separately influenced the bacterial community composition in different soil layers. In the surface soil samples, we found that the ST (F=4.421, P=0.009), SMC (F = 3.841, P = 0.021) and SOC (F = 3.403, P = 0.038) strongly influenced the bacteria community composition. In the subsurface soil samples, the SMC (F=4.016, P=0.016), DOC (F=3.755, P=0.028) and TN (F=3.435, P=0.036) were the main drivers of the bacterial community composition.

For fungi, SMC (F=5.003, P=0.004) had the strongest correlation with the community composition, followed by ST (F = 3.907, P=0.012) and DOC (F = 3.455, P=0.035). In surface soil samples, significant correlations were observed between the fungal community composition and SMC (F=4.604, P=0.008), ST (F=3.823, P=0.024) and SOC (F=3.533, P=0.034). In the subsurface, SMC (F = 4.886, P = 0.007), TN (F = 4.212, P=0.010) and DOC (F=3.854, P=0.020) were significantly correlated with soil fungal community composition.

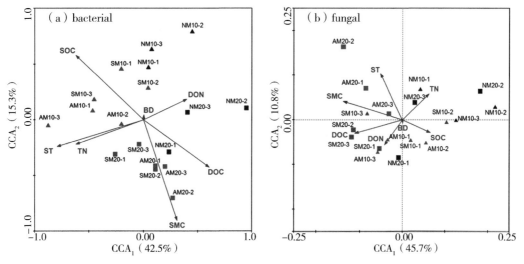

Figure 5-7 Canonical corresponding analysis (CCA) of bacterial (a) and fungal (b) community compositions with environmental variables (10, surface soil 20, subsurface soil)

5.5 Discussion

5.5.1 *Treatment effects on bacterial and fungal α-diversity*

In our research, bacterial and fungal richness and diversity were higher (exhibited as higher Chao1 and, Shannon and lower Simpson indices) in the AM and SM treatment soil samples than in the NM treatment soil samples. This finding suggested that the mulching treatments (especially the AM treatment) favored more complicated and diverse bacterial and fungal communities better than the NM treatment. Similarly, Wu et al. (2009) reported that a continuous PFM regime in paddy fields stimulated the development of diverse microbial communities better than the NM regime; these diverse microbial communities were attributed to higher amounts of high quality substrates. Moreover, it is important to recognize that the mulching treatment may increase functional diversity due to the observed increase in alpha diversity, which leads to a micro-ecological system that is more resilient to environmental change.

5.5.2 *Treatment effects on bacterial and fungal community composition*

In our research, the PCoA revealed significant differences among the three mulching treatments and clearly grouped all of the samples separately, confirming that the dominant soil microbes differed significantly in their relative abundances among different treatments. The soil in the AM and SM treatments (especially the AM treatment) had significantly higher abundances of the dominant phyla *Proteobacteria* and *Actinobacteria* than the soils from the NM treatment (Figure 5-3). Within the former phylum, the OTUs assigned to the bacterial classes *Alphaproteobacteria* were relatively frequent in both the AM and SM treatment soil samples. Similarly, Chen et al. (2014) found that the *Alphaproteobacteria* were greatly enriched in a grass mulching treatment compared with their richness in NM. As members of these classes have been shown to positively correlate with the utilization of a wide range of C compounds (Makhalanyane et al., 2015), the increased abundances of these phyla in our study may be of particular importance in biogeochemical cycling of carbon and, as such, might be suitable to address the large variety of carbon sources under mulching conditions. Additionally, we found a higher abundance of *Actinobacteria* in AM treatment soil samples than in SM and NM treatment samples (especially in surface soil). Recent studies suggest an important role for *Actinobacteria* in soil metabolic functioning due to their involvement in the decomposition of organic materials (Nielsen et al., 2014). Another striking pattern was the significant decreases in the relative abundances of the dominant phyla, *Chloroflexi* and *Firmicutes* in the AM and SM treatment soil samples. Members of *Chloroflexi* have also been found to tolerate extreme soil environments (Neilson et al., 2012). The higher frequency of *Firmicutes* in the NM treatment soil samples than in the mulching treatment soil samples may be related to their ability to produce endospores that are resistant to desiccation under NM environmental conditions (Gomez-Montano et al., 2013). Moreover, the *Firmicutes* exhibit rapid growth followed by abundant spore formation and is generally considered oligotrophic (Mueller et al., 2015).

Similar to the bacterial community, the fungal communities also responded differently to the mulching treatments, as the communities in the soil samples from the three treatments were clearly separated from each other. Moreover, *Ascomycota* and *Basidiomycota* were the most abundant fungal phyla in all of the soil samples, which are also in agreement with a previous study showing that *Ascomycota* and *Basidiomycota* accounting for more than 60% of the total sequences in soil samples as demonstrated by Illumina HiSeq sequencer (Schmidt et al., 2013). Two dominant classes within *Ascomycota* (*Eurotiomycetes* and *Sordariomycetes*) increased in relative abundance under AM conditions (Figure 5 – 4) and many more *Ascomycota* representatives contributed to the phylum-level composition shift in the surface soil samples. Regarding the abundances of the phyla in the soil samples, the enrichment of *Ascomycota* in soil samples may be associated with a higher disease suppression ability (Shen et al., 2015). The relative abundance of *Basidiomycota* among treatments presented different trends in two soil layers, with a significant increase in NM treatment surface soil samples, but a decrease in subsurface soil samples relative to the abundance in the AM treatment soil samples. This finding may suggest a higher sensitivity of this phylum to environmental changes (Acosta-Martínez et al., 2014).

For bacteria, obvious trend in our study was that *Agrobacterium* and *Pseudomonas* were significantly enhanced in the mulching treatment, and were more evident in AM treatment. Similar results were reported by Bonanomi et al. (2008), and they attributed the predominance of these bacteria in soil under mulching to the anoxic environment. Moreover, the AM treatment resulted in the significant enrichment of *Frankia* and *Nitrobacter* genera. Members of *Frankia* are known as their ability to form N_2-fixing root nodule symbioses with actinorhizal plants and *Nitrobacter* are generally considered playing an important role in the nitrogen cycle by oxidizing nitrite into nitrate in soil (Chaia et al., 2010). These findings illustrated that AM treatment has a significant ability to enhance microbes that drive soil nitrogen turnover in nutrient cycling under this agro-ecosystem. Many previous studies have demonstrated that *Corynebacterium* play important roles in degrading aromatic compounds and that *Cellulomonas* are generally associated with degrading cellulose using enzymes such as endoglucanase and exoglucanase (Akasaka et al., 2003). Additionally, *Streptomyces* can produce a number of antibiotics and other bioactive natural products that suppress soil-borne diseases (Onaka et al., 2011). Thus, the higher frequency of OTUs assigned to these three genera indicated that the AM treatment markedly increased some beneficial bacterial species compared with the other treatments. However, in contrast, OTUs assigned to the genera *Lysobacter* and *Stenotrophomonas* were significantly less abundant in the AM and SM treatment soil samples than in the NM treatment soil samples. Consistent with this finding, Hayward et al. (2010) also found that exposure to natural light and rainfall conditions increased the number of these two genera due to their ability to use lower levels of substrates than other bacterial genera more efficiently.

Among fungi, the most dominant species were assigned to the genus *Penicillium*, which includes species that have been shown to play critical roles in the production of secondary metabolites or in the decomposition of organic matter (Sarookhani and Moazzami, 2007). Obvious differences in particular fungal taxa were also apparent across the soil samples from the three treatments. Acosta-Martínez et al. (2014) found that *Penicillium* species were ubiquitous soil fungi preferring moderate climates, which may explain their increased abundance in AM and

SM treatment soil samples compared to NM. *Trichoderma*, *Talaromyces* and *Fusarium* were more frequent in AM and SM compared to those in NM treatment soil samples, and a previous study reported that members of these fungal groups were associated with degrading complex substrates, such as cellulose, chitin, and lignin-related compounds (Shen et al., 2015).

5.5.3 Links between the selected soil properties with the bacterial and fungal community composition.

The correlation analysis indicates that specific changes in bacterial and fungal communities could be partially explained by soil characteristics. In the present research, the SMC in the mulching treatment was always higher than that of NM plot in the growing season and after harvest. These results are compatible to those of previous studies which were probably due to lower run-off and evaporation under mulching (Zhang et al., 2015). The daily mean ST in the mulched plots was higher than that in the NM plots during growing season in both layers. Similar observations have also been reported by Wang et al. (2015), and they considered that the effects of the PFM on ST was related to the plastic film absorbing and reflecting solar energy. For different treatments, the AM plots had higher SMC and ST than that under the other two treatments for most sampling dates during the whole growing stage. This may be due to AM was conducted in autumn of the previous year until the harvest. Therefore, AM application could achieve seasonal adjustment of soil moisture in rainfed area by preserving rainfall, improving soil water storage before sowing, and reducing invalid water loss during fallow period. It was reported that continuous use of plastic mulch may accelerate the decomposition of the SOC (Obalum and Obi, 2010). In this study, however, the SOC did not differ significantly between different mulching treatments at the sampling date in two layers. For other properties, our results suggest that the adoption of mulching significantly increased the content of TN, DOC and DON in each layer. Similar results have been observed in other studies, e.g. in the semiarid Loess Plateau of China, and they considered that the higher moisture content arose from PFM promoting soil nutrients accumulation (Luo et al., 2015).

The stepwise regression analysis shows that ST was the most predominant factor in explaining the bacterial and fungal diversity under different mulching application (Table 5-5). Notably, ST has also been showed to have an important effect on soil bacterial and fungal community composition, as verified by the CCA model and Mantel test. This result is in general agreement with many other studies reported that temperature was the most influential factor on the soil microbial community, which were mainly due to ST could influence heterotrophic respiration by changing the activity of extracellular enzymes, the microbial respiration rate, and the substrate availability for the soil microbes (Suseela et al., 2012; Frey et al., 2008). SMC was another major factor that influences the bacterial and fungal composition under different treatments, which was supported by our observation that SMC was significantly correlated with taxonomical composition (Figure 5-7). A limited decrease in soil moisture may be a stressful process for some microorganisms, due to physical constraints that affect bacterial or fungal habitats (Or et al. ,2007). As soil drying occurs, available water in pores becomes disconnected, slowing down diffusion of solutes and limiting substrate availability resulting in a decline in nutrient flow to microbes (Schjønning et al., 2003). Apart from the ST and SMC, soil properties, such as the SOC, TN and DOC also significantly contributed to the variation in the bacterial and fungal community composition, which was probably explained by decomposing the crop residues with different chemical structure. Moreover, we found more genes and/or higher bacterial and fungal diversity in the AM than in the NM and SM, and one explanation is that mulching might cause a shift in the predominant microbial life history strategies, which favors more active, copiotrophic microbial groups (Deng et al., 2015).

5.6 Conclusion

In addition to the effects of mulching on soil properties, our research fills a gap in understanding of the effects of different mulching treatments on abundance and composition of bacterial and fungal communities in soil under maize. Our results confirmed that mulching treatments, especially AM, played an important role in improving the SMC and ST conditions throughout the growing periods, and simultaneously increased the soil nutrients (e.g. TN, DOC and DON). The α-diversity indices (Chao1, Shannon and Simpson indices) confirmed the predicted increase in bacterial and fungal diversity as a result of both AM and SM treatments (especially AM). The positive effects of AM and SM on species abundances were very similar, while the AM harbored relatively more beneficial microbial taxa than the SM, e.g., taxa related to higher degrading capacity and nutrient cycling. Canonical correspondence analysis (CCA) revealed that the soil organic carbon (SOC) content was the most predominant main factor in explaining affecting the composition of the bacterial community, while soil moisture content (SMC) was the most important factor that determined in affecting the fungal community composition. Taken together, our data indicated that AM treatment rather than SM and NM treatments is a good practice for maintaining microbial diversity and altering microbial composition. The challenge in the future will be better to understand the variation in microbial patterns over time under different mulching treatments and elucidate the expression of functional genes within each community.

5.7 Reference

Acosta-Martínez V., Cotton J., Gardner T., et al. Predominant bacterial and fungal assemblages in agricultural soils during a record drought/heat wave and linkages to enzyme activities of biogeochemical cycling [J]. Appl. Soil Ecol., 84: 69-82.

Akasaka H., Izawa T., Ueki K., et al, 2003. Phylogeny of numerically abundant culturable anaerobic bacteria associated with degradation of rice plant residue in Japanese paddy field soil [J]. FEMS Microbiol. Ecol., 43: 149-161.

Baldrian P., Kolařík M., Štursová M., et al, 2012. Active and total microbial communities in forest soil are largely different and highly stratified during decomposition [J]. Isme J., 6: 248-258.

Bao S. D, 2005. Soil and Agricultural Chemistry Analysis [M]. Beijing: Agriculture Press.

Bonanomi G., Chiurazzi M., Caporaso S., et al, 2008. Soil solarization with biodegradable materials and its impact on soil microbial communities [J]. Soil Biol. Biochem., 40: 1989-1998.

Bu L., Liu J., Zhu L., et al, 2014. Attainable yield achieved for plastic film-mulched maize in response to nitrogen deficit [J]. Eur. J. Agron., 55: 53-62.

Chaia E. E., Wall L. G., and Huss-Danell K, 2010. Life in soil by the actinorhizal root nodule endophyte Frankia. A review [J]. Symbiosis, 51: 201-226.

Chen Y., Wen X., Sun Y., et al, 2014. Mulching practices altered soil bacterial community structure and improved orchard productivity and apple quality after five growing seasons [J]. Sci. Hortic., 172: 248-257.

Cuello J. P., Hwang H. Y., Gutierrez J., et al, 2015. Impact of plastic film mulching on increasing greenhouse gas emissions in temperate upland soil during maize cultivation [J]. Appl. Soil Ecol., 91: 48-57.

Deng J., Gu Y., Zhang J., et al, 2015. Shifts of tundra bacterial and archaeal communities along a permafrost thaw gradient in Alaska [J]. Mol. Ecol., 24: 222-234.

Frey S. D., Drijber R., Smith H., et al, 2008. Microbial biomass, functional capacity, and community structure after 12 years of soil warming [J]. Soil Biol. Biochem., 40: 2904-2907.

Gomez-Montano L., Jumpponen A., Gonzales M. A., et al, 2013. Do bacterial and fungal communities in soils of the Bolivian Altiplano change under shorter fallow periods? [J]. Soil Biol. Biochem., 65: 50-59.

Hayward A. C., Fegan N., Fegan M., et al, 2010. Stenotrophomonas and Lysobacter: ubiquitous plant-associated gamma-proteobacteria of developing significance in applied microbiology [J]. J. Appl. Microbiol., 108: 756-770.

Huttenhower C., Gevers D., Knight R., et al, 2013. Structure, function and diversity of the healthy human microbiome [J]. Nature, 486: 207-214.

Kasirajan S., and Ngouajio, M, 2012. Polyethylene and biodegradable mulches for agricultural applications: a review [J]. Agron. Sustainable Dev., 32: 501-529.

Larkin R. P, 2003. Characterization of soil microbial communities under different potato cropping systems by microbial population dynamics, substrate utilization, and fatty acid profiles [J]. Soil Biol. Biochem., 35: 1451-1466.

Legendre P., and Legendre L, 2012. Numerical Ecology (3rd ed.) [M]. Elsevier: Oxford, UK.

Li F. M., Song Q. H., Jjemba P. K., et al, 2004. Dynamics of soil microbial biomass C and soil fertility in cropland mulched with plastic film in a semiarid agro-ecosystem [J]. Soil Biol. Biochem., 36: 1893-1902.

Li Z. G., et al, 2015. Plastic mulching with drip irrigation increases soil carbon stocks of natrargid soils in arid areas of northwestern china [J]. Catena, 133: 179-185.

Liu E.K., He W.Q., and Yan C.R, 2014. "White revolution" to "white pollution"-agricultural plastic film mulch in China [J]. Environ. Res. Lett., 9: 091001.

Liu Y., Mao L., He X., et al, 2012. Rapid change of AM fungal community in a rain-fed wheat field with short-term plastic film mulching practice [J]. Mycorrhiza, 22: 31-39.

Lozupone C., Lladser M. E., Dan K., et al, 2011. Unifrac: an effective distance metric for microbial community comparison [J]. Isme J., 5: 169-72.

Luo S., Zhu L., Liu J., et al, 2015. Sensitivity of soil organic carbon stocks and fractions to soil surface mulching in semiarid farmland [J]. Eur. J. Soil Biol., 67: 35-42.

Makhalanyane T. P., Valverde A., Gunnigle E., et al, 2015. Microbial ecology of hot desert edaphic systems [J]. FEMS Microbiol. Rev., 39: 203-221.

Maul J. E., Buyer J. S., Lehman R. M., et al, 2014. Microbial community structure and abundance in the rhizosphere and bulk soil of a tomato cropping system that includes cover crops [J]. Appl. Soil Ecol., 77: 42-50.

Mueller R. C., Belnap J., and Kuske C. R, 2015. Soil bacterial and fungal community responses to nitrogen addition across soil depth and microhabitat in an arid shrubland [J]. Front. Microbiol., 6: 891.

Muñoz K., Schmidt-Heydt M., Stoll D., et al, 2015. Effect of plastic mulching on mycotoxin occurrence and mycobiome abundance in soil samples from asparagus crops [J]. Mycotoxin Res., 31: 191-201.

Neilson J. W., Quade J., Ortiz M., et al, 2012. Life at the hyperarid margin: novel bacterial diversity in arid soils of the Atacama Desert, Chile [J]. Extremophiles, 16: 553-566.

Nielsen S., Minchin T., Kimber S., et al, 2014. Comparative analysis of the microbial communities in agricultural soil amended with enhanced biochars or traditional fertilisers [J]. Agric., Ecosyst. Environ., 191: 73-82.

Obalum S. E., and Obi M. E, 2010. Physical properties of a sandy loam Ultisol as affected by tillage-mulch management practices and cropping systems [J]. Soil Till. Res., 108: 30-36.

Onaka H., Mori Y., Igarashi Y., et al, 2011. Mycolic acid-containing bacteria induce natural-product biosynthesis in Streptomyces species [J]. Appl. Environ. Microbiol., 77: 400-406.

Or D., Smets B. F., Wraith J. M., et al, 2007. Physical constraints affecting bacterial habitats and activity in unsaturated porous media-a review [J]. Adv. Water Resour., 30: 1505-1527.

Prewitt L., Kang Y., Kakumanu M. L., et al, 2014. Fungal and bacterial community succession differs for three wood types during decay in a forest soil [J]. Microb. Ecol., 68: 212-221.

Sarookhani M. R., and Moazzami, N, 2007. Isolation of Acremonium species producing cephalosporine C (CPC) from forest soil in Gilan province, Iran [J]. Afr. J. Biotechnol., 6: 2506-2510.

Schjønning P., Thomsen I. K., Moldrup P., et al, 2003. Linking soil microbial activity to water and air-phase contents and diffusivities [J]. Soil Sci. Soc. Am. J., 67: 156-165.

Schmidt P. A., Bálint M., Greshake B., et al, 2013. Illumina metabarcoding of a soil fungal community [J]. Soil Biol. Biochem., 65: 128-132.

Shen Z., Ruan Y., Wang B., et al, 2015. Effect of biofertilizer for suppressing Fusarium wilt disease of banana as well as enhancing microbial and chemical properties of soil under greenhouse trial [J]. Appl. Soil Ecol., 93: 111-119.

Steenwerth K. L., Jackson L. E., Calderon F. J., et al, 2002. Soil microbial community composition and land use history in cultivated and grassland ecosystems of coastal California [J]. Soil Biol. Biochem., 35: 1599-1611.

Suseela V., Conant R. T., Wallenstein M. D., et al, 2012. Effects of soil moisture on the temperature sensitivity of heterotrophic respiration vary seasonally in an old field climate change experiment [J]. Global Change Biol., 18: 336-348.

Wang S., Luo S., Li X., et al, 2016. Effect of split application of nitrogen on nitrous oxide emissions from plastic mulching maize in the semiarid Loess Plateau [J]. Agric., Ecosyst. Environ., 220: 21-27.

Wang X., Li Z., and Xing Y, 2015. Effects of mulching and nitrogen on soil temperature, water content, nitrate-N content and maize yield in the Loess Plateau of China [J]. Agric. Water Manage., 161: 53-64.

Wang Y. P., Li X. G., Hai L., et al, 2014. Film fully-mulched ridge-furrow cropping affects soil biochemical properties and maize nutrient uptake in a rainfed semi-arid environment [J]. Soil Sci. Plant Nutr., 60: 486-498.

Wu M. Y., Wu L. H., Zhao L. M., et al, 2009. Effects of continuous plastic film mulching on paddy soil bacterial diversity [J]. Acta Agric. Scand., Sect. B., 59: 286-294.

Xie W. Y., Su J. Q., and Zhu Y. G, 2015. Phyllosphere bacterial community of floating macrophytes in paddy soil environments as revealed by illumina high-throughput sequencing [J]. Appl. Environ. Microbiol., 81: 522-532.

Zhang G. S., Hu X. B., Zhang X. X., et al, 2015. Effects of plastic mulch and crop rotation on soil physical properties in rain-fed vegetable production in the mid-Yunnan plateau, China [J]. Soil Till. Res., 145: 111-117.

Zou H. T., Zhang Y. L., Huang Y., et al, 2005. Effect of plastic mulching in previous autumn for moisture conservation on growth and yield of spring maize in semiarid region of western Liaoning Province [J]. Agric. Res. in Arid Areas, 23: 25-28.

6
Effects of Plastic Mulching on Carbon Footprint

6.1 Abstract

Producing more food with a lower environmental cost is one of the most crucial challenges worldwide. Plastic mulching has developed as one of the most dominant practices to improve crop yields, however its impacts on greenhouse gas (GHG) emissions during the production life cycle of a crop are still unclear. The objective of this work is to quantify the impacts of plastic film on GHG emissions and to reduce GHG emissions with innovative agronomy practices. Carbon footprint per unit of area (CFa), per unit of maize grain yield (CFy), and per unit of economic output (CFe) were evaluated for three maize cultivation systems-no mulch system, an annual plastic mulching system (PM) and a biennial plastic mulching pattern, namely, "one film for 2 years" system (PM2) during 2015-2018 years in a maize field located on Loess Plateau of China. Results suggested that PM induced a 24% improvement in maize yields during the four experimental years compared to a no-mulch treatment (NM). However, PM dramatically increased the CFa by 69%, 59% of which was created by the input of plastic film material, and 10% was created by increases in the soil's N_2O emissions. The yield improvements from PM cannot offset the increases in CFa, and CFy and CFe were increased by 36%. Shifting from PM to PM2 did not reduce crop yields, but it led to a 21% reduction in CFa and a 23% reduction in CFy and CFe due to the reduced input amount of plastic film, decreased soil N_2O emissions, and less diesel oil used for tillage. Compared to NM, CFy and CFe were only 5% higher in PM2. This study highlights the necessity of reducing the input amount of plastic film in the development of low-carbon agriculture and shifting from conventional PM cultivation to PM2 could be an efficient option for mitigating GHG emissions while sustaining high crop yields in plastic mulched fields.

Keywords: Carbon footprint; Greenhouse gas; Plastic film; Maize

6.2 Introduction

Food systems are responsible for 20-30% of total anthropogenic greenhouse gas (GHG) emissions (Hertwich and Peters, 2009; Vermeulen et al., 2012), and reducing GHG emissions from farmland is a critical component of climate change mitigation (Lipper et al., 2014; Tilman et al., 2011). High yields from farmland are greatly reliant on input materials and energy (e.g., fertilizer, power, and fuel); however, the production, transport, and application of agricultural inputs are usually accompanied by GHG emissions. It is estimated that the crop demand in 2050 will double (Tilman et al., 2011), and to meet the goal of insuring future food security and alleviating climate change, it is necessary to adapt agriculture technologies to achieve high crop yields with reduced GHG emissions (Chen et al., 2014).

Plastic film is an agricultural input that is widely used worldwide. The history of plastic mulching technology dates back to the mid-1950s, when Dr. Emery M. Emmert of the University of Kentucky first used polyethylene (PE) as a greenhouse film and detailed the principles of plastic technology (Emmert, 1957; Kasirajan and Ngouajio, 2012). Since then, plastic mulching practices have been adopted in many countries because of their ability to improve soil temperature, enable earlier harvests, inhibit weed growth, and conserve water (Tarara, 2000; Scarascia-Mugnozza et al., 2011; Kasirajan and Ngouajio, 2012). In China, the biggest consumer of plastic film in the world, the area of farmland using plastic mulching has increased more than 150-fold since 1982, and in 2016, this area reached 1.8×10^7 ha, which is equal to about 14% of the total farmland in China (NBSC 2017). The application of plastic film dramatically improved crop yields in China, and according to a review (Liu et al., 2014), when plastic mulching was adopted, grain and cash crop yields increased by 20–35% and 20–60%, respectively. As such, the application of plastic film makes great contributions to improving food supplies.

During the development of plastic mulching technologies, many different types of plastic mulching patterns have been created. Among them, the plastic mulching pattern called "one film for 2 years" under no-tillage conditions (PM2) was created to reduce the negative environmental effects of plastic film application (Chen et al., 2018b; Xie et al., 2018; Yan et al., 2014). The negative environmental effects of plastic film application are caused by two aspects: ①plastic residues pollution and ②increased GHG emission. PE is the most commonly used material for manufacturing plastic film, and it requires long periods to degrade in the soil (200–400 years). Once PE is incorporated into the soil, its accumulation can significantly retard crop root growth and the soil's ability to transport water and nutrients, which leads to significant yield decreases (Gao et al., 2018; Liu et al., 2014). Manufacturing plastic film is a process that requires high energy consumption, and it is estimated that the input of 1 kg of plastic film in agriculture will cause 22.72 kg of CO_2-ep emissions (Wang et al., 2017; Xue et al., 2018), which is even higher than the emission factor of N fertilizer (Chen et al., 2014; Qi et al., 2018). Therefore, the use of plastic film as an agricultural input has the potential to dramatically improve GHG emissions during a crop's production life cycle. The designed PM2 system employs a no-tillage practice to prevent incorporating plastic residue into the soil and prolongs the life of the plastic film from 1 year to 2 years in order to reduce the potential GHG emissions caused by plastic film inputs. However, it is still largely unknown how GHG emissions during a crop's production life cycle respond to different plastic mulching patterns and whether PM2 is effective in reducing GHG emissions.

Carbon footprint (CF), defined as the total amount of carbon emissions caused directly and indirectly by an activity, or the emissions that accumulate over the life cycle of a product from cradle to grave (Wiedmann and Minx, 2007), has become an important method to systematically evaluate the carbon emissions caused by artificial factors in the agricultural production process. CF evaluation for crop products provides a reliable way to recognize the key sources of

GHG emissions in different cultivation systems and to quantify the influences of agricultural practices and other environmental factors (Chen et al., 2014; Gan et al., 2014). As such, it provides the theoretical basis for optimizing agricultural management practices to reduce GHG emissions and develop low-carbon agriculture. In this study, maize (*Zea mays* L.) was chosen as the test crop because it is one of the three largest food staples cultivated worldwide (FAO, 2018). Previous research has investigated the CF response in maize production in terms of fertilizer management practices (Grassini and Cassman, 2012; Ha et al., 2015; Jat et al., 2019; Ma et al., 2012; Qi et al., 2018; Wang et al., 2015), irrigation practices (Grassini and Cassman, 2012; Zhang et al., 2018), tillage practices (Zhang et al., 2013; Zhang et al., 2016), diversified crop rotation systems (Ma et al., 2012; Yang et al., 2014), and the spatial variation of CF on regional and national scales (Cheng et al., 2015; Xu and Lan, 2017; Zhang et al., 2017). While in previous researches that focused on the responses of carbon footprint to plastic mulching, He et al. (2018) investigated the influences of plastic mulch technology on carbon footprint on national scale, and Xue et al. (2018) reported the response of carbon footprint to different plastic mulching patterns in different winter wheat production systems. However, little has been reported about the influences of different plastic mulching patterns on the CF of maize production.

The objective of this study is to evaluate the CF response to different plastic mulching patterns in a maize field and to fill the knowledge gaps in developing low-carbon crop production in fields with plastic mulching. Based on this, it was hypothesized that ①the application of plastic mulching can lead to significant improvements in field GHG emissions, and ②shifting from a conventional plastic mulching pattern (PM) to PM2 is an effective way to achieve the goals of improving crop yields while mitigating GHG emissions. To test these hypotheses, the CF was evaluated based on four years of field experiments for three cultivation systems: ①the no-mulch system (NM); ②a PM system with annual film replacement; and ③the PM2 system.

6.3 Materials and methods

6.3.1 *Site description*

The field experiment was carried out in the Shouyang rain-fed agricultural experimental station (37°45′N, 113°12′E, 1080 m altitude) in the Shanxi Province of Northern China from 2015 to 2018. The research area is characterized by a temperate semiarid continental monsoon climate with four distinct seasons, and the average annual air temperature, rainfall amount, and frost-free period are 7.4 ℃, 480 mm, and 140 days during the past 48 years (year 1967-2014) according to statistics from local meteorological bureau. During the experimental year 2015-2018, rainfall data acquired form experiment observations was 386 mm, 461 mm, 482 mm and 362 mm, respectively. The soil texture is sandy loam, and the soil is classified as a Calcaric Cambisol, according to the world reference base for soil resources (FAO, 2006). Prior to the beginning of the experiment, the upper 20 cm of the soil had a pH of 7.8, a soil organic matter content of 18.03 g·kg^{-1}, and a total N of 0.85 g·kg^{-1}. Spring maize (*Zea mays* L.) is the main crop cultivated in the research area; it is usually sown in late April or early May and harvested in late September or early October. After harvest, there is a fallow period until sowing the following year.

6.3.2 *Experimental design and management*

Three treatments were selected from the research site for this study: ①the NM system, in which the maize was cultivated without the application of plastic film; ②the PM pattern; and ③the new plastic mulching pattern, PM2 ("one film for 2 years" under no-tillage conditions), which was designed to reduce residual pollution from plastic and the CF of the plastic mulching field. During the four experimental years, there were two cycles for PM2 (i.e., the first cycle in 2015–2016 and the second cycle in 2017–2018). The experiment employed a completely randomized design with three replicates, and each plot area measured 60 m^2 (6 m× 10 m).

In NM, rotary tillage was carried out to a depth of 30 cm, and in accordance with local practice, fertilizers were applied at rates of N 225 kg·ha^{-1} (urea), P_2O_5 162 kg·ha^{-1} (calcium superphosphate), and K_2O 45 kg·ha^{-1} (potassium chloride) before sowing. Spring maize was sown at a row distance of 60 cm and plant spacing of 30 cm (i.e., a sowing density of 55,556 plants per ha). The maize cultivar "Qiangsheng 51" was sown in early May for the four experimental years. After seedling emergence, seedling thinning was carried out manually, and herbicide was used to control weeds. The maize was harvested in early October, and maize grain drying and threshing were completed manually. All of the maize stubble was removed from the field manually for subsequent use as animal fodder.

In PM, PE plastic film with a width of 80 cm and a thickness of 10 μm was used. Between two strips covered by the plastic film, there was a bare strip; the width of bare strips were 40 cm, which led to soil coverage of about 67%. The same tillage and fertilization methods used in NM were adopted for PM, and after fertilization, the maize was sown directly into the film using a hole-sowing tool, with same row distance and plant spacing to NM (60 cm and 30 cm). After harvest, following local practices, the film was not recycled but was incorporated into the soil during tillage, forming so-called "white pollution" (Liu et al., 2014).

In PM2, a type of PE-based plastic film withanti-ageing additives (a mixture of hindered amine light stabilizer, an antioxidant, and ultraviolet-absorbent materials, about 3% of total weight of plastic film) was used, and the production of this kind of plastic film did not need to change the manufacture processes. Tillage was carried out in the first year to form a relatively smooth soil surface and reduce potential damage to the film during the film-laying process. In the subsequent three years, a no-tillage practice was adopted to prevent the potential incorporation of plastic residue into the soil. For the first growing season with PM2, the same sowing, fertilization, and tillage methods used in NM and PM were applied, and in the second growing season, to prevent further film destruction, the maize was sown into bare soil near the edge of the film, and fertilizer was only applied in the middle of the bare strip. The film in PM2 was used continuously for 2 years, and it was removed manually and replaced with a new film prior to the sowing of the third growing season. The fertilization amount, seedling thinning, weed control, plant residue management, sowing, and harvest times were same in NM, PM, and PM2.

6.3.3 *Field measurement*

6.3.3.1 Maize yield

The central 4 rows in each plot were selected to measure maize yield at harvest in 2015–2018. For each row, the center 5 m was hand-harvested, the fresh weight and water content of the grains were measured, and then the grain yield per plot was calculated on a "wet-mass basis" (standard water content of 15.5%) (Payero et al., 2008).

6.3.3.2 Measurement of N_2O emissions

The static closed chamber method was applied to measure the N_2O emissions in different treatments (Hutchinson and Mosier, 1981). The static chamber comprised an anchor with a hollow ring attached to its brim, and a chamber with an inner diameter of 24 cm and a height of 25 cm. During measurement, the hollow ring on the brim was filled with water to seal the chamber while the gas was sampled. A fan mounted inside the chamber was used to mix the head-space air, and a temperature recorder mounted inside the chamber was used to record the air temperature. In NM, the anchors were installed in the middle of the maize rows, and in PM and PM2, they were installed in the middle of the bare strip and mulched strip to account for the potential permeation of N_2O through the mulch films (Figure 6-1, Berger et al., 2013; Nishimura et al., 2012; Nishimura et al., 2014). Gases were collected between 9:00 and 11:00 am weekly during the growing season of 2015–2018, and a sampling syringe with a volume of 50 ml was used for gas sampling at 0, 20, and 40 min after chamber closure. N_2O concentrations in the gas samples were analyzed with a gas chromatograph equipped with an electron capture detector and a Poropak Q column (Agilent 7890A, Agilent Technologies, Inc., Santa Clara, USA). The N_2O emission rates were calculated from the linear increase in N_2O concentrations per unit of surface area of the chamber for a specific time interval, and the cumulative N_2O emissions were estimated using linear interpolation, as reported by Cuello et al. (2015).

6.3.4 *Calculation of CF*

In this study, the system boundary for CF calculation was defined as the entire production life

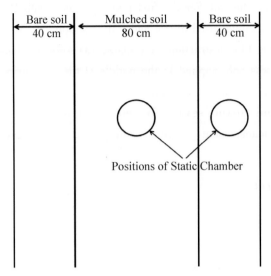

Figure 6-1 Scheme of positions for gas sampling with static chamber methods in plastic mulched field

cycle, from raw material exploitation and the production and transportation of agricultural inputs (e.g., plastic film, fertilizer, diesel, and pesticides) to the farm gates (maize harvest). In accordance with previous research on CF evaluation for maize fields (Qi et al., 2018; Samarappuli and Berti, 2018), three factors for the sources of GHG emission were included in this study: ① the manufacture, storage, and transportation of the agriculture inputs, including the plastic film, fertilizers, pesticides, and seeds to the farm gates as well as their application; ② the energy consumption of machinery operation, including tillage, sowing, and harvest; and ③ field N_2O emissions resulting from N fertilizer application. Soil CO_2 flux was not taken into account because the net flux was much less than gross CO_2 emissions that can be measured, and net fluxes usually contributed less than 1% to the global warming potential of agriculture on a global scale (Chen et al., 2014). CH_4 emission may be negligible because of its small percentage in GHG emissions in drylands (Guo and Zhou, 2007; Xue et al., 2018; Ceschia et al., 2010; Huang et al., 2019), and human labor was not taken into account because humans respired CO_2 regardless of whether or not they were working (West and Marland, 2002). Such a CF evaluation method was in accordance with most previous researches that were carried in short-term field studies (Xue et al., 2018; Wang et al., 2017; Qi et al., 2018; Samarappuli and Berti, 2018; Liu et al., 2016). The indirect total amount of GHG emissions associated with the agricultural inputs (CF_{inputs}), the CF per unit of area (CFa), per unit of maize grain yield (CFy), and per unit economic output (CFe) expressed in equal amounts of CO_2 (CO_2-ep) were calculated as:

$$CF_{inputs} = \sum_i (Q_{usedi} \times \delta_i) \quad (6-1)$$

$$CFa = CF_{inputs} + CF_{N_2O} \quad (6-2)$$

$$CF_{N_2O} = N_2O_{cumu} \times 44/28 \times 265 \quad (6-3)$$

$$CFy = CFa/\text{Yield} \quad (6-4)$$

$$CFe = CFa/TE \quad (6-5)$$

Where the Q_{usedi} is the used amount for the ith type of individual agricultural input (kg·year^{-1}); δ_i is the emissions factor for the ith type of individual agricultural input (kg CO$_2$-eq·kg^{-1}); CF_{N_2O} is the CF associated with soil N$_2$O emissions (kg CO$_2$-ep·year^{-1}), N_2O_{cumu} is the cumulative N$_2$O emissions (kg NO$_2$-N·ha^{-1}·year^{-1}), and the global warming potential of N$_2$O relative to CO$_2$ over 100 years was assumed to be 265 (IPCC, 2014); yield is the measured maize grain yield (kg·year^{-1}); TE is the total economic output of the crop (CNY·ha^{-1}). Average maize grain prices from 2015 to 2018 year was used for the calculation of TE, and it was 1.71 CNY·kg^{-1} (NDRC, 2019).

In this study, tillage was completed with a small walking tractor, and sowing and harvest were completed manually; however, to stay in accordance with actual production, we used the surveyed oil consumption data in actual production near the experimental site for the CF calculation, and the inputted amount of diesel was converted from L·ha^{-1} to kg·ha^{-1} with a density value of 0.86 kg·L^{-1} (Samarappuli and Berti, 2018). The amount of plastic film used in PM and PM2 (Q_{film}, kg·ha^{-1}) was calculated as (Chen et al., 2018b):

$$Q_{film} = F_{mulch} \times Thick_{film} \times \rho_{film} \times 10000 \quad (6-6)$$

Where F_{mulch} stands for the ratio of area covered by plastic film (67% in PM and PM2), $Thick_{film}$ and ρ_{film} were the thickness and density of the plastic film (0.01 mm and 930 kg·m^{-3}). Therefore, the input amount of the plastic film was calculated as 63.0 kg·ha^{-1} in the PM and PM2 treatments. As additives used in the plastic film of PM2 only took up a small proportion of total weight of plastic film, same calculation methods were adopted for PM and PM2.

The emissions factors for different inputted materials were obtained from IKE eBalance V3.0 (IKE Environmental Technology CO., Ltd., China; Qi et al., 2018; Wang et al., 2017), and the emissions factors for the seeds, urea, calcium superphosphate, potassium chloride, diesel oil, herbicide, and plastic film were 0.58, 2.30, 0.30, 0.13, 0.89, 10.15, and 22.72 kg CO$_2$-ep·kg^{-1}, respectively. The total economic output of the crop was calculated as the maize grain yield multiplied by its price (Huang et al., 2019; Yang et al., 2014).

6.3.5 Statistical analysis

The effects of PM and PM2 on crop yields and N$_2$O emissions were analyzed with one-way analysis of variance (ANOVA) using SAS v8.0 software (SAS Institute, Cary, NC, USA), and least significant differences (LSD) were used to detect the mean differences between the treatments.

6.4 Results and discussion

6.4.1 *The response of maize yield to different plastic mulching patterns*

As shown in Figure 6-2, compared to NM, maize yields were significantly improved by 12%, 25%, 31%, and 29% in PM ($P<0.05$), and significantly improved by 12%, 20%, 35%, and 38% in PM2 for 2015, 2016, 2017, and 2018 ($P<0.05$), respectively. PM and PM2 significantly increased the four years average yield by of 24% and 26%, respectively and no significant difference in maize yields was found for PM and PM2.

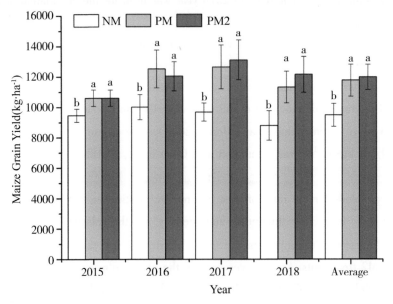

Figure 6-2 Maize yield in the no mulch (NM), conventional plastic mulching (PM) and "one film for 2 years" plastic mulching (PM2) cultivation systems. Different letters (a, b) in each column indicate significant differences ($P<0.05$) between treatments for each experimental year and average value of four experimental years according to LSD tests

These results were in accordance with previous researches which suggested that plastic mulching was an effective practice by which to improve crop yields due to its improvements of soil temperature and moisture conditions (Gao et al., 2018; Li et al., 2013; Zhang et al., 2011). Zhang et al. (2011) reported that maize yields were improved by 8-24% under different plastic mulching systems. Li et al. (2013) suggested that maize yields with plastic mulching were increased by 13.0%. The meta-analysis from Gao et al. (2018) using 266 peer-reviewed publications in China showed that plastic mulching increased crop yields by about 24%. The increased yields mainly resulted in reduced evaporation under the plastic film which reduced the ratio of evaporation to transpiration, leaving more water available for plant uptake (Ding et al., 2013; Li et al., 2008). Furthermore, the application of plastic film has been reported to improve soil temperature conditions by preventing the loss of emitted radiant energy in the long-wave spectrum, establishing an insulating air gap, and reducing latent heat loss caused by evaporation (Ham et al., 1993; Ham et al., 1994). As such, it can reduce water and temperature stress for crop growth and contribute to yield improvements.

Shifting from PM to PM2 did not induce any significant crop decline. This can be attributed to the prolonged lifetime of the plastic film. The lifetime of the plastic film is greatly influenced by the oxidation process, which occurs through photo-, thermal-, or γ-initiated oxidation (Basfar and Ali, 2006); however, plastic film stability can be improved by several to a dozen times with the application of a hindered amine light stabilizer, an antioxidant, and ultraviolet-absorbent materials (Al-Salem, 2009; Basfar and Ali, 2006). In our previous study, only slight damage was observed on the plastic film in the second year of PM2, but it did not result in significant influences on the soil moisture and temperature conditions compared to PM (Chen et al., 2018b). These results are in accordance with the reports of Xie et al., (2018) and Xu et al. (2013). In Xie et al. (2018), it was found that PM2 can significantly improve maize yields compared to NM treatments, and in Xu et al. (2013), it was suggested that PM2 has similar improvement effects on maize yields compared to other plastic mulching patterns.

6.4.2 Response of N_2O emissions to different plastic mulching patterns

In this study, the measured yearly N_2O emissions ranged from 1.56–2.65 kg $N_2O-N \cdot ha^{-1}$, which equaled 0.69% to 1.18% of the applied nitrogen fertilizer (Figure 6-3). These values are comparable to the IPCC (2013), which suggests that direct N_2O emissions from agriculture are estimated to be equal to 1% of the applied nitrogen, with an uncertainty range of 0.3–3.0%.

Compared to NM, PM induced significantly higher N_2O emissions over the four experimental years. In the PM treatment, the N_2O emissions from the mulched strip accounted for 34–40% of the total N_2O emissions, and 60–66% of the N_2O emissions were produced from the bare strip. This is in accordance with previous reports suggesting that N_2O can permeate the mulch film and that it plays an important role in N_2O emissions in plastic mulched fields (Berger et al., 2013; Nishimura et al., 2012; Nishimura et al., 2014). In Nishimura et al. (2014), the contribution of N_2O emissions through the permeation of the mulch film was found to range from 33–53% of the total emissions in plastic mulched fields. The increase in N_2O emissions with plastic mulching may be attributed to a higher availability of mineral N, reduced soil aeration, and improved soil temperature and moisture conditions (Cuello et al., 2015; Nishimura et al., 2012; Zhang et al., 2018). Nitrification and denitrification are two biochemical processes responsible for N_2O emissions. These processes are largely influenced by

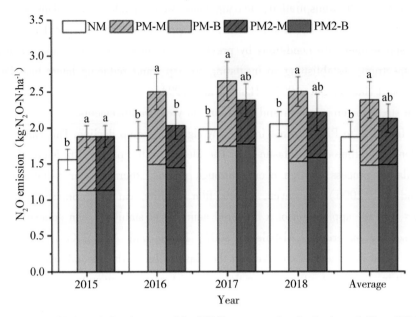

Figure 6-3 N$_2$O emission in no mulch (NM), conventional plastic mulching (PM) and "one film for 2 years" plastic mulching (PM2) cultivation systems. N$_2$O emission in PM and PM2 is comprised of emission from the bare strip (B) and the mulched strip (M), and total N$_2$O emission is the sum of B and M. Error bars stand for the standard errors and different letters over error bars stand for the significant differences between treatments at $P<0.05$

the availability of mineral N (Das and Adhya, 2014; Pajares and Bohannan, 2016) and are highly sensitive to soil aeration (Barnard et al., 2005; Gillam et al., 2008). The application of plastic mulching has been shown to improve the mineral N availability in soil due to improved soil temperature and moisture (Tian et al., 2013; Zhang et al., 2012). These enhanced conditions along with a reduction in the gas exchange between the soil and the air, lead to a higher N$_2$O emission rate. As such, in most studies, it has been confirmed that the application of plastic mulching could increase N$_2$O emissions (Cuello et al., 2015; Nan et al., 2016; Nishimura et al., 2012; Zhang et al., 2018). However, responses of soil N$_2$O emission to plastic mulching are also highly depend on the soil and climate conditions. In two consecutive wheat-maize rotation cycles, Chen et al. (2017) observed annual N$_2$O emission decreased in the field with plastic mulching compared to no mulch field in one cycle, while increased in another cycle. Berger et al. (2013) observed plastic mulching decreased N$_2$O emission in sandy soils because sandy soil was usually difficult to maintain soil moisture.

Compared to PM, N$_2$O emissions were found to be significantly lower in PM2 in 2016 and also slightly lower in the other experimental years. Previous research suggested that the effects of plastic mulching on soil moisture and temperature generally declined from the middle of the mulched strip to the center of the bare strip (Chen et al., 2018a; Chen et al., 2019). Accordingly, in PM2, when fertilizer was only applied to the bare strip, it effectively avoided the effects of higher moisture and temperature on N_2O emissions caused by plastic mulching.

6.4.3 GHG emissions associated with agricultural inputs

In the NM, PM, and PM2 systems, GHG emissions associated with agricultural inputs (CF_{inputs}) was 1563, 3001, and 2249 kg CO_2-eq·ha^{-1}·$year^{-1}$, respectively (Table 6-1). The range of CF_{inputs} in the current study was lower than that of Qi et al. (2018), in which maize was cultivated with irrigation and larger fertilizer load, and it was similar to that of Shi et al. (2011) in which maize was cultivated with medium amounts of fertilizer. The differences in CF_{inputs} in different studies can be explained by the requirements for input materials under different management practices.

In NM, urea accounted for 72% of CF_{inputs} and was the largest contributor to CF_{inputs}, which is in accordance with previous research carried out in NM maize fields (Chen et al., 2014; Qi et al., 2018). In PM, most CF_{inputs} came from the input of plastic film and urea, which were 48% and 37% of GHG emissions associated with agricultural inputs, respectively. The calculated GHG emissions associated with the input of plastic film (1,438 kg CO_2-eq·ha^{-1}·$year^{-1}$) were similar to the values reported in a wheat field (1,022-2,726 kg CO_2-eq·ha^{-1}·$year^{-1}$; Xue et al., 2018). In PM2, the input of plastic film and urea contributed to 32% and 50% of the CF_{inputs}. Across treatments, diesel oil contributed 3-6% of the CF_{inputs} (accounting for tillage, sowing and harvest process), with the lowest contribution to the PM2 treatment where tillage was carried out only in the first year, and no-tillage practice was employed for the other three years. Compared to NM, CF_{inputs} was increased by 92% and 44% in PM and PM2, respectively (Table 6-1), as a result of the input of plastic film. Compared to PM, CF_{inputs} was reduced by 21% because of the lower input amount of plastic film and the reduction in diesel oil for tillage.

Table 6-1 Greenhouse gas (GHG) emission caused by input of materials in no mulch (NM), conventional plastic mulching (PM) and "one film for 2 years" plastic mulching (PM2) cultivation systems

Agricultural inputs	Input amount (kg·ha^{-1}·$year^{-1}$)			GHG emission (kg CO_2-eq·ha^{-1}·$year^{-1}$)		
	NM	PM	PM2	NM	PM	PM2
Urea	489	489	489	1,125	1,125	1,125
Calcium Superphosphate	953	953	953	286	286	286
Potassium chloride	82	82	82	11	11	11
Diesel oil-Tillage	48	48	12	43	43	11
Diesel oil-Sowing	29	29	29	26	26	26
Diesel oil-Harvest	39	39	39	34	34	34
Herbicide	2	2	2	20	20	20

(continued)

Agricultural inputs	Input amount (kg·ha⁻¹·year⁻¹)			GHG emission (kg CO_2-eq·ha⁻¹·year⁻¹)		
	NM	PM	PM2	NM	PM	PM2
Plastic film	0	63	32	0	1,438	719
Total	-	-	-	1,563	3,001	2,249

6.4.4 Carbon footprint

During the experimental years, the CF per unit of area (CFa) was 2,341, 3,993 and 3,134 kg CO_2-eq·ha⁻¹·year⁻¹, the CF per unit of maize grain yield (CFy) was 0.25, 0.34 and 0.26 kg CO_2-eq·kg⁻¹, and the CF per unit of economic output (CFe) was 0.14, 0.20 and 0.15 kg CO_2-eq·CNY⁻¹ for NM, PM, and PM2, respectively (Figure 6-4). The urea input and soil N_2O emissions were the two most important sources of CF in the NM system, and contributed 48% and 33% of the CFa, respectively. In the PM system, the most important source of GHG emissions was the input of plastic film (36% of the total GHG emissions), which was followed by the input of urea (28%) and soil N_2O emissions (25%). In the PM2 system, the input of urea, plastic film, and soil N_2O emissions accounted for 36%, 23% and 28% of CFa, respectively. Compared to NM, CFa was 70% higher in PM, of which 61% was contributed by the input of plastic film, and 9% was contributed by higher soil N_2O emissions. Although crop yield was improved by PM, it failed to offset the huge increases in CFa, and this led to a 37% higher CFy. Shifting from PM to PM2 reduced the CFa by 22% as a result of the reduction in the input of plastic film, N_2O emissions, and diesel for tillage. These reductions led to PM2 system having a CFy value that is only 6% higher than that of the NM.

As maize is a commonly cultivated crop worldwide, the CF for its production has been investigated widely. In a project comparing the CF under different cultivation systems in 153 site field experiments in China, it was suggested that the CFa for maize ranged from 2,273-8,269 kg CO_2-eq·ha⁻¹, and the CFy ranged from 0.287-1.179 kg CO_2-eq·kg⁻¹ (Chen et al., 2014). In an irrigated maize field located in central Nebraska, Grassini and Cassman (2012) reported CFy as 0.23 kg of CO_2-eq·kg⁻¹. In a rain-fed maize field located in Ottawa, Canada, Ma et al. (2012) reported that the CFa ranged from 243-5,284 kg CO_2-eq·ha⁻¹ in different fertilization and crop rotation systems. Using farm survey data from eastern China, Yan et al. (2015) estimated the CFa and CFy in this region were (2.3±0.1) t CO_2-eq·ha⁻¹ and (0.33 ± 0.02) kg CO_2-eq·kg⁻¹, respectively. Although large variations in CF have been found among the different cases due to variations in environmental conditions and cultivation methods, most of those cases suggest that N fertilizer is the biggest source of CF in fields with plastic mulching. This was confirmed by the NM treatment of this study. However, our study

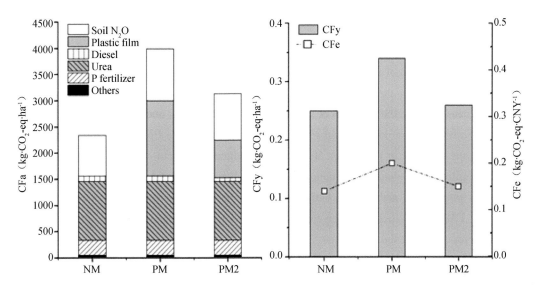

Figure 6-4 Carbon footprint per unit of area (CFa), carbon footprint per unite of maize grain yield (CFy) and carbon footprint per unit of economic output (CFe) in no mulch (NM), conventional plastic mulching (PM) and "one film for 2 years" plastic mulching (PM2) cultivation systems. Greenhous gas emission caused by soil N_2O emission, input of plastic film, diesel, urea, phosphate fertilizer and other materials (seeds, herbicide and potassium chloride) were taken into account for the calculation of carbon footprint. Presented data is the average of the experimental years 2015-2018

suggests that when plastic film is applied in maize production, the total CF is dramatically improved, and the biggest source of total GHG emissions shifts from the input of N fertilizer to the input of plastic film. Considering the demand for food caused by a growing population (Tilman et al., 2011), it is hard to stop the use of plastic film, especially in regions where crop production is limited by water or temperature stress. As such, this study suggested that shifting from PM to PM2 was a promising method to improve maize yield meanwhile reduce the carbon footprint in maize production system with plastic mulching.

In addition, it should been noted that the CF was estimated based on four years' field experiments in this study, and the soil carbon sequestration was not considered for the CF evaluation because relative short experimental periods. However, no-tillage practice which was employed in PM2 had been widely recognized as a kind of agricultural practices to improve the soil carbon sequestration in long-term running (Lu et al., 2009; Baker et al., 2007), and long-term application of PM2 might result in lower carbon footprint. Moreover, accumulation of plastic residue in PM had been proven to lower the crop yield in long-term running (Gao et al., 2018; Zhang et al., 2020), which might further improve the CFy and CFe in PM. For those, further long-term observations were needed to be carried out in the future.

6.5 Conclusions

Although the application of plastic film in conventional patterns can improve crop yields, it led to dramatic increases on the area-scale and yield-scale CF as a result of indirect greenhouse emissions caused by the input of plastic film materials and increased soil N_2O emissions. Shifting from a conventional plastic mulching pattern to a PM2 cultivation system did not lead to any declines in yield, but it effectively reduced the yield-scale CF to levels similar to the NM cultivation system. It was concluded that reducing indirect GHG emissions caused by the input of plastic film materials is extremely important to control the CF of plastic mulched fields, and PM2 is an efficient option for achieving the goals of mitigating GHG emissions and improving crop yields at the same time.

6.6 Reference

Al-Salem S M, 2009. Influence of natural and accelerated weathering on various formulations of linear low density polyethylene (LLDPE) films [J]. Materials and Design, 30 (5), 1729-1736.

BAKER J M., OCHSNER T E, VENTEREA R T, et al., 2007. Tillage and soil carbon sequestration-What do we really know? [J]. Agriculture, ecosystems & environment, 118, 1-5.

Barnard R, Leadley P W, Hungate B A, 2005. Global change, nitrification, and denitrification: a review [J]. Global Biogeochemical Cycles, 19, GB1007.

Basfar A A, Ali K I, 2006. Natural weathering test for films of various formulations of low density polyethylene (LDPE) and linear low density polyethylene (LLDPE) [J]. Polymer Degradation and Stability, 91 (3), 437-443.

Berger S, Kim Y, Kettering J, et al., 2013. Plastic mulching in agriculture-Friend or foe of N_2O emissions? [J] Agriculture Ecosystems and Environment, 167, 43-51.

Ceschia E, Béziat P, Dejoux J F, et al, 2010. Management effects on net ecosystem carbon and GHG budgets at European crop sites [J]. Agriculture, Ecosystems & Environment, 139, 363-383.

Chen B, Garré S, Liu H, et al., 2019. Two-dimensional monitoring of soil water content in fields with plastic mulching using electrical resistivity tomography [J]. Computers and Electronics in Agriculture, 159, 84-91.

Chen B, Liu E, Mei X, et al., 2018a. Modelling soil water dynamic in rain-fed spring maize field with plastic mulching [J]. Agricultural water management, 198, 19-27.

Chen B, Yan C, Garré S, et al., 2018b. Effects of a "one film for 2 years" system on the grain yield, water use efficiency and cost-benefit balance in dryland spring maize

(Zea mays L) on the Loess Plateau, China [J]. Archives of Agronomy and Soil Science, 64 (7), 939-952.

Chen H, Liu J, Zhang A, et al, 2017. Effects of straw and plastic film mulching on greenhouse gas emissions in Loess Plateau, China: a field study of 2 consecutive wheat-maize rotation cycles [J]. Science of the Total Environment, 579, 814-824.

Chen X, Cui Z, Fan M, et al, 2014. Producing more grain with lower environmental costs [J]. Nature, 514 (7523), 486.

Cheng K, Yan M, Nayak D, et al., 2015. Carbon footprint of crop production in China: an analysis of National Statistics data [J]. The Journal of Agricultural Science, 153 (3), 422-431.

Cuello J P, Hwang H Y, Gutierrez J, et al., 2015. Impact of plastic film mulching on increasing greenhouse gas emissions in temperate upland soil during maize cultivation [J]. Applied Soil Ecology, 91, 48-57.

Das S, and Adhya T K, 2014. Effect of combine application of organic manure and inorganic fertilizer on methane and nitrous oxide emissions from a tropical flooded soil planted to rice [J]. Geoderma, 213, 185-192.

Ding R, Kang S, Zhang Y, et al., 2013. Partitioning evapotranspiration into soil evaporation and transpiration using a modified dual crop coefficient model in irrigated maize field with ground-mulching [J]. Agricultural water management, 127, 85-96.

Emmert E M, 1957. Black polyethylene for mulching vegetables [J]. Proceedings American Society for Horticultural Science, 69, 464-469.

FAO (Food and Agricultural Organization). 2018. FAOSAT, Statistical Databases, United Nations, Rome. http://www.fao.org/faostat/en/#data/QC.

FAO (Food and Agriculture Organization). 2006. Guidelines for soil description [M]. 4th edn, Rome.

Gan Y, Liang C, Chai Q, et al., 2014. Improving farming practices reduces the carbon footprint of spring wheat production [J]. Nature Communications, 5, 5012.

Gao H, Yan C, Liu Q, et al., 2018. Effects of plastic mulching and plastic residue on agricultural production: A meta-analysis [J]. Science of the Total Environment, 651, 484-492.

Gillam K M, Zebarth B J, and Burton D L, 2008. Nitrous oxide emissions from denitrification and the partitioning of gaseous losses as affected by nitrate and carbon addition and soil aeration [J]. Canadian Journal of Soil Science, 88, 133-143.

Grassini P, and Cassman K G, 2012. High-yield maize with large net energy yield and small global warming intensity [J]. Proceedings of the National Academy of Sciences, 109 (4), 1074-1079.

Guo J, and Zhou C, 2007. Greenhouse gas emissions and mitigation measures in Chinese agroecosystems [J]. Agricultural and Forest Meteorology, 142, 270-277.

Ha N, Feike T, Back H, et al., 2015. The effect of simple nitrogen fertilizer recommendation strategies on product carbon footprint and gross margin of wheat and maize production in the north china plain [J]. Journal of Environmental Management, 163, 146-154.

Ham J M, Kluitenberg G J, Lamont W J, 1993. Optical properties of plastic mulches affect the field temperature regime [J]. Journal of the American Society for Horticultural Science, 118 (2), 188-193.

Ham J M and Kluitenberg G J, 1994. Modeling the effect of mulch optical properties and mulch-soil contact resistance on soil heating under plastic mulch culture [J]. Agricultural and Forest Meteorology, 71 (3-4), 403-424.

He G, Wang Z, Li S, et al., 2018. Plastic mulch: Tradeoffs between productivity and greenhouse gas emissions [J]. Journal of Cleaner Production, 172, 1311-1318.

Hertwich E G, and Peters G P, 2009. Carbon footprint of nations: A global, trade-linked analysis [J]. Environmental Science and Technology, 43 (16), 6414-6420.

Huang J, Chen Y, Pan J, et al., 2019. Carbon footprint of different agricultural systems in China estimated by different evaluation metrics [J]. Journal of Cleaner Production, 225, 939-948.

Hutchinson G L, and Mosier A. R, 1981. Improved soil cover method for fifield measurement of nitrous oxide fluxes [J]. Soil Science Society of America Journal, 45, 311-316.

IPCC (Intergovernmental Panel on Climate Change). 2013. T F Stocker, D Qin, G K Plattner, M M B Tignor, S K Allen, J Boschung, A Nauels, Y Xia, V Bex, P M Midgley (Eds.), Climate Change 2013: The Physical Science Basis. Contribution of Working Group I to the Fifth Assessment Report of the Intergovernmental Panel on Climate Change, Cambridge University Press, Cambridge, United Kingdom and New York, NY, USA.

IPCC (Intergovernmental Panel on Climate Change). 2014. Core Writing Team, R K Pachauri, L A Meyer (Eds.), Climate Change 2014: Synthesis Report. Contribution of Working Groups I, II and III to the Fifth Assessment Report of the Intergovernmental Panel on Climate Change, IPCC, Geneva, Switzerland.

Jat S L, Parihar C M, Singh A K, et al., 2019. Energy auditing and carbon footprint under long-term conservation agriculture-based intensive maize systems with diverse inorganic nitrogen management options [J]. Science of The Total Environment, 664, 659-668.

Kasirajan S, and Ngouajio M, 2012. Polyethylene and biodegradable mulches for agricultural applications: a review [J]. Agronomy for Sustainable Development, 32, 501-529.

Li R, Hou X Q, Jia Z K, et al., 2012. Effects of rainfall harvesting and mulching technologies on soil water, temperature, and maize yield in Loess Plateau region of China [J]. Soil Research, 50 (2), 105-113.

Li R, Hou X, Jia Z, et al., 2013. Effects on soil temperature, moisture, and maize yield of cultivation with ridge and furrow mulching in the rainfed area of the Loess Plateau China [J]. Agricultural Water Management, 116, 101-109.

Li S, Kang S, Li F, et al., 2008. Evapotranspiration and crop coefficient of spring maize with plastic mulch using eddy covariance in northwest China [J]. Agricultural Water Management, 95 (11), 1214-1222.

Lipper L, Thornton P, Campbell B M, et al., 2014. Climate-smart agriculture for food security [J]. Nature climate change, 4 (12), 1068.

Liu E K, He W Q, and Yan C R, 2014. "White revolution" to "white pollution"-agricultural plastic film mulch in China [J]. Environmental Research Letters (9): 3.

Liu Q, Liu B, Ambus P, et al, 2016. Carbon footprint of rice production under biochar amendment-a case study in a Chinese rice cropping system [J]. Global Change Biology, 8, 148-159.

LU F, Wang X, Han B, et al., 2009. Soil carbon sequestrations by nitrogen fertilizer application, straw return and no-tillage in China's cropland [J]. Global Change Biology, 15, 281-305.

Ma B L, Liang B C, Biswas D K, et al., 2012. The carbon footprint of maize production as affected by nitrogen fertilizer and maize-legume rotations [J]. Nutrient Cycling in Agroecosystems, 94 (1), 15-31.

Nan W G, Yue S C, Huang H Z, et al., 2016. Effects of plastic film mulching on soil greenhouse gases (CO_2, CH_4 and N_2O) concentration within soil profiles in maize fields on the Loess Plateau, China [J]. Journal of integrative agriculture, 15 (2), 451-464.

NBSC (National Bureau of Statistics China). 2017. China Rural Statistical Yearbook [M]. Beijing: China Statistics Press.

NDRC (National Development and Reform Commission). 2019. Compilation of national cost and income of agricultural products [M]. Beijing: China Statistics Press. (in Chinese).

Nishimura S, Komada M, Takebe M, et al., 2014. Contribution of nitrous oxide emission from soil covered with plastic mulch film in vegetable field [J]. Journal of Agricultural Meteorology, 70, 117-125.

Nishimura S, Komada M, Takebe M, et al., 2012. Nitrous oxide evolved from soil covered with plastic mulch film in horticultural field. Biology and Fertility of Soils, 48, 787-795.

Pajares S, and Bohannan B J. M, 2016. Ecology of nitrogen fixing, nitrifying, and denitrifying microorganisms in tropical forest soils [J]. Frontiers in Microbiology, 7, 1045.

Payero J O, Tarkalson D D, Irmak S, et al., 2008. Effect of irrigation amounts applied with subsurface drip irrigation on corn evapotranspiration, yield, water use efficiency, and dry matter production in a semiarid climate [J]. Agricultural water management, 95, 895-908.

Qi J Y, Yang S T, Xue J F, et al., 2018. Response of carbon footprint of spring maize production to cultivation patterns in the Loess Plateau, China [J]. Journal of Cleaner Production, 187, 525-536.

Samarappuli D, and Berti M T, 2018. Intercropping forage sorghum with maize is a promising alternative to maize silage for biogas production [J]. Journal of Cleaner Production, 194, 515-524.

Scarascia-Mugnozza G, Sica C, and Russo G, 2011. Plastic materials in european agriculture: actual use and perspectives [J]. Journal of Agricultural Engineering, 42, 15-28.

Shi L G, Chen F, Kong F L, et al., 2011. The carbon footprint of winter wheat-summer maize cropping pattern on North China Plain [J]. China Population, Resources and Environment., 21, 93-98.

Tarara J M, 2000. Microclimate modification with plastic mulch [J]. HortScience, 35, 169-180.

Tian J, Lu Q, Fan M, et al., 2013. Labile soil organic matter fractions as influenced by non-flooded mulching cultivation and cropping season in rice-wheat rotation [J]. European Journal of Soil Biology, 56, 19-25.

Tilman D, Balzer C, Hill J, et al., 2011. Global food demand and the sustainable intensification of agriculture [J]. Proceedings of the National Academy of Sciences, 108 (50), 20260-20264.

Tilsner J, Wrage N, Lauf J, et al., 2003. Emission of gaseous nitrogen oxides from an extensively managed grassland in NE Bavaria, Germany-II Stable isotope natural abundance of N_2O [J]. Biogeochemistry, 63, 249-267.

Vermeulen S J, Campbell B M, and Ingram J S I, 2012. Climate change and food systems [J]. Annual Review of Environment and Resources, 37, 195-222.

Wang Z B, Wen X Y, Zhang H L, et al., 2015. Net energy yield and carbon footprint of summer corn under different N fertilizer rates in the north china plain [J]. Journal of Integrative Agriculture, 14 (8), 1534-1541.

Wang Z B, Chen J, Mao S C, et al., 2017. Comparison of greenhouse gas emissions of chemical fertilizer types in China's crop production. Journal of Cleaner Production, 141, 1267-1274.

West T, and Marland G, 2002. Net carbon flux from agricultural ecosystems: methodology for full carbon cycle analyses [J]. Environment Pollution, 116, 439-444.

Wiedmann T, and Minx J, 2007. A definition of "carbon footprint" [J]. Ecological Economics Research Trends, 2, 55-65.

Xie J H, Zhang R Z, Li L L, et al., 2018. Effects of plastic film mulching patterns on maize grain yield, water use efficiency, and soil water balance in the farming system with one film used two years [J]. The Journal of Applied Ecology, 29 (6), 1935-1942.

Xu W Q, Yang Q F, Niu F J, et al., 2013. Effects of stalk returned to the field and film mulching on the soil physical and chemical characteristics and the maize growth [J]. Journal of Maize Sciences, 21 (3), 87-93.

Xu X and Lan Y, 2017. Spatial and temporal patterns of carbon footprints of grain crops in china [J]. Journal of Cleaner Production, 146, 218-227.

Xue J F, Yuan Y Q, Zhang H L, et al., 2018. Carbon footprint of dryland winter wheat under film mulching during summer-fallow season and sowing method on the Loess Plateau [J]. Ecological indicators, 95, 12-20.

Yan C R, He W Q and Neil C, 2014. Plastic-film mulch in Chinese agriculture: importance and problems [J]. World Agriculture, 4 (2), 32-36.

Yan M, Cheng K, Luo T, et al., 2015. Carbon footprint of grain crop production in China-ased on farm survey data [J]. Journal of Cleaner Production, 104, 130-138.

Yang X, Gao W, Zhang M, et al., 2014. Reducing agricultural carbon footprint through diversified crop rotation systems in the North China Plain [J]. Journal of Cleaner Production, 76, 131-139.

Zhang D, Ng E L, Hu W, et al., 2020. plastic pollution in croplands threatens long-term food security [J]. Global Change Biology. DOI: 10.1111/gcb.15043.

Zhang D, Shen J, Zhang F, et al., 2017. Carbon footprint of grain production in China [J]. Scientific reports, 7 (1), 4126.

Zhang M Y, Wang F J, Chen F, et al., 2013. Comparison of three tillage systems in the wheat-maize system on carbon sequestration in the north china plain [J]. Journal of Cleaner Production, 54, 101-107.

Zhang S, Li P, Yang X, et al., 2011. Effects of tillage and plastic mulch on soil water, growth and yield of spring-sown maize [J]. Soil and Tillage Research, 112 (1), 92-97.

Zhang W, He X, Zhang Z, et al., 2018. Carbon footprint assessment for irrigated and rainfed maize (Zea mays L.) production on the Loess plateau of China [J]. Biosystems Engineering, 167, 75-86.

Zhang X Q, Pu C, Zhao X, et al., 2016. Tillage effects on carbon footprint and ecosystem services of climate regulation in a winter wheat-summer maize cropping system of the north china plain [J]. Ecological Indicators, 67, 821-829.

Zhang X, Bai W, Gilliam F S, et al., 2011. Effects of in situ freezing on soil net nitrogen mineralization and net nitrification in fertilized grassland of northern china [J]. Grass and Forage Science, 66 (3), 391-401.

Zhang X, Wang Q, Gilliam F S, et al., 2012. Effect of nitrogen fertilization on net nitrogen mineralization in a grassland soil, northern China [J]. Grass and Forage Science, 67 (2), 219-230.

Zhou L M, Li F M, Jin S L, et al., 2009. How two ridges and the furrow mulched with plastic film affect soil water, soil temperature and yield of maize on the semiarid Loess Plateau of China [J]. Field Crops Research, 113 (1), 41-47.